Land and Climate

Cover and interior design: © Agence Coam

ISBN: 978-2-493291-15-8

© 2021 Éditions La Butineuse
Atelier des Entreprises
Place de l'Europe – Porte Océane 3
56400 Auray – France
www.editions-labutineuse.com

Patrick Love

Land and Climate

Insights from the IPCC Special Report

Table of contents

Note to readers

The aim of this book is to enable people without the necessary time or background to better understand an IPCC report published in 2019.

The title of this report is *Climate Change and Land: An IPCC Special Report on climate change, desertification, land degradation, sustainable land management, food security, and greenhouse gas fluxes in terrestrial ecosystems.*[1] The report is also known as the *Special Report on Climate Change and Land.*

The Special Report "assesses the dynamics of the land-climate system, and the economic and social dimensions of addressing the challenges of land degradation, desertification and food security in a changing climate. It also assesses the options for governance and decision-making across multiple scales". The Special Report was compiled by over 600 experts from varying fields of research. The majority of them (52%) are from developing countries.

The findings are based on over 7000 scientific and technical publications. Each finding is grounded in an evaluation of underlying evidence and agreement. A level of confidence is expressed using five qualifiers: very low, low, medium, high and very high.

In this short introduction, we only cite findings that are classed as "high confidence" or "very high confidence". These are quoted word for word in our text, and are highlighted in italics.

The following text follows the structure of the IPCC report:

Chapter 1 introduces land-climate interactions at local, regional and global scales.

Chapter 2 examines the forces driving desertification and land degradation, and how these relate to human activity and climate change.

Chapter 3 describes the causes of desertification, and how these causes interact with climate change.

Chapter 4 focuses on food security, and the impacts of climate change on food systems, considering how mitigation and adaptation can contribute to both human and planetary health.

Chapter 5 analyses the interlinkages between options for climate mitigation and adaptation to address desertification and land degradation, and to enhance food security.

The concluding chapter assesses the opportunities, decision making and policy responses to risks in the climate-land-human system.

Since the examples and discussions in our text draw on sources other than the IPCC, the opinions expressed may not be those of the IPCC or its members.

Notes

1. *IPCC, 2019: Climate Change and Land: an IPCC special report on climate change, desertification, land degradation, sustainable land management, food security, and greenhouse gas fluxes in terrestrial ecosystems* [P.R. Shukla, J. Skea, E. Calvo Buendia, V. Masson-Delmotte, H.-O.Pörtner, D. C. Roberts, P. Zhai, R. Slade, S. Connors, R. van Diemen, M. Ferrat, E. Haughey, S. Luz, S. Neogi, M. Pathak, J. Petzold, J. Portugal Pereira, P. Vyas, E. Huntley, K. Kissick, M. Belkacemi, J. Malley, (eds.)]. In press. https://www.ipcc.ch/srccl/

Chapter One
Land-climate interactions

The term 'butterfly effect" comes from the work of meteorologist Edward Lorenz, who discovered in 1961 that a tiny change in data he typed into his numerical weather model (rounding up 0.506127 to 0.506) eventually led to a totally different forecast from the one using the original number.[1] The idea that the flap of a butterfly's wings in Brazil could lead to a tornado in Texas is far from Newton's "For every action, there is an equal and opposite reaction". In a Newtonian world, the flap of the wings would only cause a reaction in the air the wings pushed, allowing the butterfly to fly. But in talking about the climate, we have to move beyond action-reaction pairs to look at complex series of interactions involving the air, the land, the oceans, and human activity. We can view these interactions as an adaptive system, where each part is constantly influencing the others and being influenced by them.

In this chapter, we will look at one part of the dynamics of this system, land-climate interactions at local, regional and global scales. Land cover and land use are adapted to "climate envelopes", combinations of ranges of temperature and/or rainfall. Greenhouse gas (GHG) emissions caused by humans impact land through changes in the weather and climate and by modifying the composition of the atmosphere, especially by increasing the amount of CO_2.

Anthropogenic warming has resulted in shifts of climate zones, primarily as an increase in dry climates and decrease of polar climates. Ongoing warming is projected to result in new, hot climates in tropical regions and to shift climate zones poleward in the mid- to high latitudes and upward in regions of higher elevation. (IPCC SRCCL, p. 44)

Energy moves from warmer regions to colder regions – from the tropics to the poles – and as the amount of heat in a region increases, the characteristics of that region change. The Sahara Desert for example grew by 10% during the 20th century,[2] while Antarctic ice is now melting six times faster than in the 1990s.[3]

The shift of warmer climate envelopes into high latitude areas could help agriculture through extended growing seasons, warmer seasonal temperatures and increased atmospheric CO_2 concentrations which boost photosynthesis. However, plants and animals have evolved to adapt to their climate niches. Even if they could change as much in the next few decades as they changed over the past thousands or millions of years, that still might not be fast enough to cope with changes projected to occur by the 2070s.[4] Overall, loss of vegetation productivity in many parts of the world could overwhelm any benefits to land use and land cover from increased atmospheric CO_2 concentrations.

Warming will also increase snowmelt and reduce albedo, the amount of solar radiation reflected rather than absorbed by a surface. When albedo is reduced, the land absorbs more heat. In polar regions, this is leading to a feedback loop where the increased heat absorption melts more ice, reducing albedo further. A similar phenomenon is happening in the tundra, where permafrost melting is releasing GHG trapped in the soil.

A counter-example from Spain shows how changes in land conditions from human use, literally from greenhouses, can affect climate. The semi-arid Almeria province has both Europe's biggest desert and the world's biggest area of greenhouses, nearly 30,000 ha. The white plastic sheeting the greenhouses are made of has increased albedo so much that the average annual temperature of the region has actually cooled by almost one degree since the 1980s.[5]

The human species developed within a climate niche. For the past 6000 years, most of us have lived in areas where the average annual temperature is 13°C, but over the next 50 years, 1 to 3 billion people could be living in areas hotter than that.[6] Moreover, some areas are already very close to experiencing combinations of heat and humidity that make it impossible to survive for long outside, for example the southern Persian Gulf shoreline and northern South Asia, home to millions of people.[7]

> *The frequency and intensity of some extreme weather and climate events have increased as a consequence of global warming and will continue to increase under medium and high emission scenarios. [...] future climate variability expected to enhance the risk and severity of wildfires in many biomes such as tropical rainforests.* (p. 45)

Flooding as an existential threat to life is mentioned in the Legend of Gilgamesh and later in the Bible, which also talks about destruction by fire. These archaic fears are coming to pass more and more frequently. The number of floods and other hydrological events has increased four-fold worldwide since 1980 and doubled since 2004. Extreme temperatures, droughts, and forest fires have more than doubled since 1980, as have storms.

As the summer of 2021 showed, all parts of the world are affected. In July alone, floods and the landslides they caused killed over 920 people, including 219 in Belgium and Germany; 192 in Mumbai and Maharashtra, India; 113 in Nuristan Province, Afghanistan; and 99 in Henan Province, China.[8] Nine people were killed by wildfires in California, but deaths related to environmental disasters are not always immediate. Over 33,000 deaths a year can be attributed to air pollution from wildfires, including 7,000 in Japan; over 3,000 in Mexico; more than 1,200 in China; more than 5,200 in South Africa; nearly 5,300 in Thailand; and almost 3,200 in the US.[9]

In 2019, the worst natural disaster worldwide in terms of human lives lost was a heatwave in Europe, responsible for 2500 of the 11,755 deaths caused by natural disasters reported that year.[10] Globally, heatwaves have become more frequent and longer since the 1950s, and the biggest increases are seen in areas that are expected to suffer most from the impacts of climate change.[11] By the end of the century, heatwaves may become extremely long (more than 60 consecutive days) and frequent (once every two years) in most areas of the world.

Droughts too are expected to get worse, with some studies predicting that by 2100, climate change could reduce terrestrial water storage (TWS) in many regions independently of other factors such as land use changes, especially in the Southern Hemisphere, the United States and southwestern Europe. The global land area and population in extreme-to-exceptional TWS drought could more than double, increasing from 3% to 7% for land area and 8% for population.[12]

There is an economic price to pay for natural disasters as well. In the US for example, the cost of 285 major weather and climate disasters since 1980 is estimated at over $1.875 trillion.[13]

The shift in climate zones means that wildfires are also shifting. In "traditional" wildfire regions, life has had time to adapt to the fires, through flame-resistance in tree bark for instance. Some species even rely on fires, like the giant California redwoods, that benefit from fires clearing undergrowth and enabling their seeds to germinate. But we are now seeing huge fires in new areas, or areas where they were very rare. The northern part of the planet is warming more quickly than the Earth as a whole, and boreal (northern) forests are now burning at a rate unseen for the past 10,000 years.[14]

Fire damage to tropical rainforests is getting worse too. Often, these fires are started deliberately to clear land for agriculture, but climate plays a role as well. Warming of the North Atlantic and tropical east Pacific oceans draws moisture away from the Amazon, making the dry season longer and increasing the risk that fires started deliberately will spread out of control.[15]

Changes in land conditions modulate the likelihood, intensity and duration of many extreme events. (p. 47) *Regional climate change can be dampened or enhanced by changes in local land cover and land use but this depends on the location and the season.* (p. 135)

Land conditions also affect "normal" climate and weather. Given the lack of historical data, there is no direct observation of how past land use changes affected atmospheric dynamics and physics globally or regionally, so models are used to estimate trends. However, climate modelling experiments only assess the impacts of land cover changes such as deforestation or urbanisation, and neglect the effects of changes in land management such as irrigation, use of fertilizers, or choice of crops. Because of this, we will use the term 'land cover changes'.

Human-induced changes to land cover and land use are part of a cycle. Deforestation and afforestation, grazing, irrigation, urbanisation, and so on cause changes in atmospheric CO_2. This provokes changes in global atmospheric variables (temperature, precipitation, atmospheric circulation, *et cetera*) which alter local and regional variables such as temperature, precipitation, and wind. These in turn lead to changes in land functioning, land cover, and structure, including photosynthesis, drying, greening, distribution of natural ecosystems, and species composition. These changes then alter atmospheric CO_2.

In boreal regions, where projected climate change will migrate the treeline northward, increase the growing season length, and thaw permafrost, regional winter warming will be enhanced by decreased surface albedo and snow. Warming will be dampened during the growing season due to larger evapotranspiration (evaporation of water from the land surface plus transpiration from plants). In the tropics, wherever climate change increases rainfall, vegetation growth and associated increase in evapotranspiration will dampen regional warming.

Agriculture, forestry and other land use (AFOLU) is a significant net source of GHG emissions. (p. 45)

Historical changes in anthropogenic land cover have resulted in a mean annual global warming of surface air from biogeochemical effects such as carbon emissions, despite being offset to some extent by biophysical cooling, from increased surface albedo as seen in Almeria for instance. The key question is how much AFOLU contributes to global anthropogenic GHG emissions, starting the cycle described above.

Agriculture, forestry and other land use were responsible for around 23% of anthropogenic GHG emissions over 2007-2016, and were the main anthropogenic source of nitrous oxide (N2O) from fertilizer, the most important GHG after carbon dioxide and methane. Chemical fertilizers using nitrogen boost agricultural production, but plants can only use a certain amount of nitrogen. After this limit is reached in croplands, N2O emissions grow exponentially. Global N2O emissions have increased over the past two decades and the fastest growth has been since 2009, notably in China and Brazil due to increased use of nitrogen fertilizers and the expansion of nitrogen-fixing crops such as soybean.[16]

The impacts of changes to land cover are not restricted to the area where they occur. Changes in local land cover or available water from irrigation can affect climate in regions hundreds of kilometres away. This is one reason why torrential rain in Germany may be linked to urbanisation in Spain. Rainfall on the Spanish coast comes mainly from Mediterranean sea breezes that pick up extra moisture from marshes and wetlands. But as the marshes and wetlands are built on or drained for farming, the breezes cannot not pick up enough moisture for summer storms over the mountains. The water vapour they had picked up is returned to the Mediterranean, where it accumulates before drifting northward and falling as rain when it cools over central Europe.[17]

Climate change is already affecting the health and energy demand of large numbers of people living in urban areas. (p. 188)

Over half the world's population live in towns and cities, generating around three-quarters of global total carbon emissions from energy use. The proportion of urban population is predicted to reach about 70% by the middle of this century, with most of the growth in the developing world. Of the world's 33 megacities in 2018 (population of 10 million or more), 27 are located in the less developed regions, and 9 of the 10 cities projected to become megacities between 2018 and 2030 are located in developing countries.[18]

Urban air pollution increases with climate change because warmer air can damage air quality by enhancing the formation of photochemical oxidants such as N2O and increasing the concentration of air pollutants such as ozone. In 2015, around 8.8 million people died as a consequence of air pollution, compared with 7.2 million from smoking tobacco. Without fossil fuel emissions, global mean life expectancy would increase by 1.1 years, and by 1.7 years if all potentially controllable anthropogenic emissions were removed.[19] Some air pollution is "natural", from dust storms for example. Data from ice cores shows that dust storms worsened during past global heating periods, and the same is likely to happen again.[20]

Urban sprawl is projected to consume an extra 1.8% to 2.4% of current cultivated land by 2030 and 5% by 2050, with highly productive lands experiencing the highest rate of conversion to urban areas landscapes. These expanding urban areas will experience direct and indirect climate change impacts, such as sea level rise and storm surges, increasing soil salinity, and landslides caused by extreme rain storms. Badly planned urbanisation can increase people's exposure to climate hazards, because informal settlements and poorly-built infrastructure are often the most exposed to fire, flooding and landslides. By 2050, population growth in flood-prone lands, climate change, deforestation, loss of wetlands, and rising sea levels are expected to increase the number of people vulnerable to flood disaster to 2 billion.[21]

Several mitigation response options have technical potential for
>3 GtCO$_2$-eq yr^{-1} by 2050 through reduced emissions and Carbon
Dioxide Removal (CDR), some of which compete for land and other
resources, while others may reduce the demand for land. (p. 49)

Emissions from AFOLU could be reduced significantly by changing certain practices, preferences and behaviours. Reduced deforestation and forest degradation could reduce emissions by the equivalent of over 0.7 to 0.8

Gigatons of CO_2 a year ($GtCO_2$-eq yr^{-1}). The total forest area in the world is 4.06 billion hectares, a reduction of 420 million hectares since 1990, through conversion to other land uses. Agricultural expansion continues to be the main driver of deforestation and forest degradation and the associated loss of forest biodiversity. Large-scale commercial agriculture (primarily cattle ranching and cultivation of soya beans and palm oil) accounted for 40% of tropical deforestation between 2000 and 2010, and local subsistence agriculture for another 33%. The rate of deforestation has decreased though, to 10 million hectares per year in 2015 to 2020, compared with 16 million hectares per year in the 1990s.[22]

A shift towards plant-based diets could save 0.7-8.0 $GtCO_2$-eq yr^{-1}, but while most people in a global survey conducted by the UN supported protection of land and forests (54%), investment in renewable energy (53%) and climate-friendly agriculture (52%), only 30% agreed that switching to a plant-based diet is a good way to combat climate change.[23]

One of the most effective potential means of cutting emissions would also save money: reducing food and agricultural waste (0.8-4.5 $GtCO_2$-eq yr^{-1}). Around a third of the food produced in the world for human consumption every year – approximately 1.3 billion tonnes – is lost or wasted, and nearly half of all fruit and vegetables produced globally are never eaten by humans. This costs roughly $680 billion in industrialised countries and $310 billion in developing countries. Apart from the benefits to the environment, saving even a quarter of the food currently lost or wasted globally would be enough to feed 870 million hungry people.[24] (See Chapter 4, *Food Security*, for a more detailed study).

Large-scale implementation of mitigation response options that limit warming to 1.5°C or 2°C would require conversion of large areas of land for afforestation/reforestation and bioenergy crops, which could lead to short-term carbon losses. (p. 49)

The change of global forest area needed to mitigate climate change varies from about -0.2 to +7.2 Mkm^2 between 2010 and 2100, depending and the assumptions of the models and scenarios used, while land demand for bioenergy crops ranges from about 3.2 to 6.6 Mkm^2 in 2100. In high-carbon lands such as forests and peatlands, the carbon benefits of land protection are greater in the short-term than converting land to bioenergy crops for carbon capture and storage, which can take several harvest cycles to 'payback' the carbon emitted during conversion.

Sustainable land management requires a combination of approaches, but the key to most practices is increasing soil organic carbon (SOC) stocks. Besides contributing to climate change mitigation by removing CO_2 from the atmosphere, enhancing SOC improves soil health and fertility, water and nutrient retention capacity, food production potential, and resilience to drought, although the scale of potential benefits depends on local conditions.

Large-scale adoption of sustainable land management practices in all croplands, grazing lands, forests and woodlands could sequester about 1-2 Gt Carbon per year globally within 30-50 years in theory. However, since the rate of sequestration declines as the soil approaches saturation level, the main carbon sequestration potential is in degraded soils. In soils that already have high SOC content, preventing losses is the priority. Overall, sustainable land management provides an opportunity to recover between 21 to 51 Gt of the lost carbon in the world's agricultural and degraded soils.[25]

Conclusion

In this chapter, we looked at the natural system and its dynamics, the impacts of climate change on land, and the feedbacks arising from biogeochemical and biophysical exchange fluxes. In the next chapter, we will examine the urgency of tackling degradation across all land ecosystems, and consider the consequences of inaction.

Notes

1. E. N. Lorenz, (1993) *The Essence of Chaos*, U. Washington Press, Seattle

2. https://journals.ametsoc.org/view/journals/clim/31/9/jcli-d-17-0187.1.xml

3. https://www.nature.com/articles/s41586-018-0179-y

4. Rates of change in climatic niches in plant and animal populations are much slower than projected climate change | Proceedings of the Royal Society B: Biological Sciences (royalsocietypublishing.org)

5. Surface temperature cooling trends and negative radiative forcing due to land use change toward greenhouse farming in southeastern Spain - Campra - 2008 - Journal of Geophysical Research: Atmospheres - Wiley Online Library

6. Future of the human climate niche | PNAS

7. The emergence of heat and humidity too severe for human tolerance | Science Advances (sciencemag.org)

8. Worldwide – Over 920 People Killed in Floods and Landslides in July 2021

9. Mortality risk attributable to wildfire-related PM2·5 pollution: a global time series study in 749 locations

10. https://cred.be/sites/default/files/adsr_2019.pdf

11. Increasing trends in regional heatwaves | Nature Communications

12. Global terrestrial water storage and drought severity under climate change | Nature Climate Change

13. Billion-Dollar Weather and Climate Disasters: Time Series | National Centers for Environmental Information (NCEI) (noaa.gov)

14. Recent burning of boreal forests exceeds fire regime limits of the past 10,000 years | PNAS

15. Amazon Basin climate under global warming: the role of the sea surface temperature | Philosophical Transactions of the Royal Society B: Biological Sciences (royalsocietypublishing.org)

16. Acceleration of global N 2 O emissions seen from two decades of atmospheric inversion | Nature Climate Change

17. Extreme hydrometeorological events and climate change predictions in Europe - NASA/ADS (harvard.edu)

18. World Urbanization Prospects - Population Division - United Nations

19. Loss of life expectancy from air pollution compared to other risk factors: a worldwide perspective | Cardiovascular Research | Oxford Academic (oup. com)

20. A 2000 Year Saharan Dust Event Proxy Record from an Ice Core in the European Alps - Clifford - 2019 - Journal of Geophysical Research: Atmospheres - Wiley Online Library

21. 2012 - 4th Edition | United Nations World Water Development Report (unesco.org)

22. State of the World's Forests 2020 (fao.org)

23. The Peoples' Climate Vote | UNDP

24. Worldwide food waste | ThinkEatSave (unep.org)

25. UNCCD_Report_Sustainable Land Management contribution to successful land-based climate change adaptation and mitigation SLM_web_ v2.pdf

Chapter Two
Land degradation

Today, people forced to leave their homes because of drought or some other "natural" disaster are called environmental migrants or refugees. In 1930s America, they were called Okies, because so many of them were driven from their homes in the Oklahoma Dust Bowl created by a combination of drought and wind that "blew away everything but your debts". This was a natural disaster in the sense that it came after a series of severe droughts and dust storms hit the US and Canadian prairies from 1930 to 1936. But drought wasn't unusual in these regions, nor were high winds, and the prairies had resisted similar episodes in the past.

The difference was that a few exceptional periods of higher than average rainfall had encouraged extensive and intensive farming, with no attempt to protect the soil. People believed that "rainfall follows the plow".[1] It doesn't, and when the grasses that held the soil in place and trapped moisture had been ploughed under, the droughts turned the earth to dust that the wind then carried away. So-called black blizzards blotted out the sun and caused drastic, rapid drops in temperature. Millions of acres of farmland were lost, and hundreds of thousands of people made destitute.

Recovery from the Dust Bowl required coordinated government intervention under the New Deal, involving legislation on soil conservation and erosion mitigation, programmes to buy livestock and other produce from affected areas, and subsidies to farmers, as well as the return of rainfall to pre-drought levels. When the next drought cycle hit in the 1950s, there was no repeat of the Dust Bowl.

Today, desertification and land degradation continue, often in places that do not have the financial and other means the US had to tackle the problem. In this chapter we will examine the forces driving desertification and land degradation, and how these relate to human activity and climate change.

Land degradation is defined as a negative trend in land condition, caused by direct or indirect human-induced processes including anthropogenic climate change, expressed as long-term reduction or loss of at least one of the following: biological productivity, ecological integrity, or value to humans.
(IPCC SRCCL, p. 53)

Apart from producing food and fibres, the land provides other services, including regulating microclimates, storing carbon and water. A report by a research coalition led by the United Nations University estimates that loss of these ecosystem services because of land degradation costs $6.3 trillion to $10.6 trillion annually, or the equivalent of 10-17% of global GDP.[2]

The International Union for Conservation of Nature (IUCN) estimates that soils store more carbon than the planet's biomass and atmosphere combined. If better land management increased the carbon stocks in the top metre of soils by just 1%, this would be higher than the amount of carbon in the CO_2 produced annually by burning fossil fuels[3]. On the contrary, when land is degraded, soil carbon can be released into the atmosphere, along with nitrous oxide, contributing to climate change.

The Intergovernmental Science-Policy Platform on Biodiversity and Ecosystem Services (IPBES) estimates that over 43% of the planet's land is degraded through human activity, negatively impacting the well-being of at least 3.2 billion people.[4]

Climate change exacerbates the rate and magnitude of several ongoing land degradation processes and introduces new degradation patterns. [...] Land-use changes and unsustainable land management are direct human causes of land degradation, with agriculture being a dominant sector driving degradation. (p. 53)

Land can be degraded by human activity, natural causes, or a combination of both. For example, 80% of global deforestation has been attributed to agricultural practices, in many cases encouraged by government subsidies worth $200 billion annually.[5] A report to the September 2021 UN Food systems Summit calculated that almost 90% of the $540bn annual global subsidies given to farmers fuels the climate crisis, damages people's health, destroys nature, and drives inequality by excluding smallholder, many of whom are women or from marginalised communities.[6] The IPBES estimates that over 70% of all natural, ice-free land has been transformed by human activity, and this could rise to 90% by 2050 if today's global land use patterns continue. If policies to limit global warming stay the same, yearly soil loss could increase by 66% globally by 2070 compared to today, reaching over 70 billion tons of soil a year.[7]

Land use, whether for agriculture, towns, or other activities, has adapted to the climate "envelopes" mentioned in Chapter 1 – combinations of temperature, humidity, *et cetera*. As the Earth warms, these envelopes shift towards the poles and towards higher altitudes. In the shorter term, this may benefit the regions they shift to, by making their growing seasons longer and enhancing photosynthesis due to higher CO_2 concentrations. But there is a price to pay. Snowmelt will increase, and permafrost will start to melt, releasing more methane and CO_2. The earth's albedo will be reduced, meaning less sunlight will be reflected back, thereby increasing warming. In the Himalayan region, 129 million farmers depend on the annual melt of snow and glaciers for irrigation. As faster melting reduces the stored water available in future years, a range of services from ecosystems to hydropower will suffer. Increased melt is already causing dangerous flash floods.[8]

Other envelopes will shift too, with the tropics seeing increases in the frequency, intensity and duration of extreme events such as heatwaves, intense rainfall, drought, and storms.[9] Farming systems will no longer be adapted, with negative impacts on crop productivity and irrigation needs. Loss of vegetation productivity in many parts of the world would be greater than any benefits to land use and land cover from increased atmospheric CO_2 concentrations. Every 1°C of warming above historical levels could cause a decrease of approximately 5% in crop productivity, making it harder to achieve the 60% increase in food production needed by 2050 to cope with population increases and changing diets.[10]

Land degradation is a driver of climate change through emission of greenhouse gases (GHGs) and reduced rates of carbon uptake. [...] Land degradation and climate change, both individually and in combination, have profound implications for natural resource-based livelihood systems and societal groups. (p. 53)

Trees and plants are the most well-known natural means of capturing carbon from the atmosphere, absorbing a third to a fourth of CO_2 emissions from fossil fuels.[11] However, the carbon they store is usually released sooner or later. This can happen over geological timescales as when plants from the carboniferous period over 300 million years ago were slowly transformed into coal to be used as fuel. The period's name comes from the Latin "carbo" – coal, and "fero" – I bear. The process happens sooner when trees are burned in wildfires or deliberately. The Amazon rainforest used to be a carbon sink, but land degradation due to clearing the trees to make way for soy and cattle production means that the Amazon is now actually a net contributor to greenhouse gas emissions.[12]

A vicious circle is created where fewer trees means less rain and higher temperatures, increasing the risk and severity of droughts and wildfires.

Farming communities and others whose livelihoods depend on natural resources have been adapting and innovating for millennia. But when change happens quickly or on a large scale, the coping strategies themselves can cause long-term harm after their short-term benefits are exhausted.

Many different social and economic domains can be affected. In Zimbabwe, livelihoods are highly-dependent on rain-fed agriculture, making them susceptible to climate-related risks. When disaster strikes, households may sell crucial assets like animals or tools, or over exploit forest products. They may cut food consumption, damaging their health, or behave in ways that destroy social cohesion, such as stealing.[13]

The economic price to pay for the loss of biosystem services can be high. To return to the Amazon, when the forest is cleared, the sudden rise in temperatures has been calculated to cost the Brazilian soy industry £3.5 billion a year, expected to rise as global warming gets worse.[14] Extreme-heat regulation services could be a longer-term source of income for owners of ecosystems. Income diversification schemes like this have been shown to help alleviate rural poverty,[15] but the owners of the medium and large-size farms that dominate Brazilian agriculture may prefer the immediate gains from soy production.

Large-scale implementation of dedicated biomass production for bioenergy increases competition for land with potentially serious consequences for food security and land degradation. (p. 53)

Other diversification strategies that were initially seen as "win-win" for the environment and the rural economy are now facing criticism, notably bioenergy programmes. Bioenergy accounts for a tenth of world total primary energy supply today. Its contribution to final energy demand is five times higher than wind and solar photovoltaic combined, even excluding traditional uses of biomass such as burning wood for domestic heating and cooking. Bioenergy's share is expected to increase as part of strategies to meet commitments to reduce fossil use. But growing crops for energy takes land away from food production. In 2009 already, a study by the Earth Policy Institute showed that the grain grown by US farmers to make biofuels was enough to feed 330 million people at average world consumption rates.[16]

In that same year, the EU's Renewable Energy Directive classified biomass as a renewable energy source,[17] encouraging national governments to promote biomass and driving up demand for wood. Burning wood for fuel releases more carbon than burning coal,[18] and in February 2021, 500 scientists wrote to the presidents of United States, the EU, Japan and South Korea asking them to stop subsidies to wood burning.[19]

The farming practices associated with growing crops for fuel also cause concern. The International Union for the Conservation of Nature (IUCN) argues that intensive and monocultural bioenergy production can cause nutrient losses and land degradation.[20] The IUCN is also concerned that the impact will be worse if homogeneous genetically improved species are favoured due to their productivity, while their effects on biodiversity, land health and ecosystem services are overlooked. The negative impacts of monoculture on land have several causes. Monocultures generally require high inputs of chemical fertilizer and pesticides that exhaust the soil further and often require intensive irrigation that increases salinization. Once again, this creates a vicious circle where natural nutrients and other bioservices are destroyed, requiring more artificial inputs. There could be a case for using land that is already degraded for biomass production, but the area of suitable land is highly uncertain, and cannot be established without considering current land use and ownership.

The IPCC concludes that reducing unsustainable use of traditional biomass reduces land degradation and emissions of CO_2, while providing social and economic benefits.

Land degradation can be avoided, reduced or reversed by implementing sustainable land management, restoration and rehabilitation practices that simultaneously provide many co-benefits, including adaptation to and mitigation of climate change. (p. 55)

The aim of the UN Decade on Ecosystem Restoration (UNDER) 2021-2030 being co-organised with WOCAT (World Overview of Conservation Approaches and Technologies) is to massively scale up the restoration of degraded ecosystems to fight climate change; enhance food security, water supply and biodiversity; and manage associated risks of conflict and migration.[21] Sustainable land management (SLM) is at the core of this programme. According to UN estimates, the restoration of 350 million hectares of degraded terrestrial and aquatic ecosystems during the Decade could generate $9 trillion in ecosystem services. Restoration could also remove 13 to 26 gigatons of greenhouse gases from the atmosphere. "The economic benefits of such interventions exceed nine times the cost of investment, whereas inaction is at least three times more costly than ecosystem restoration" according to the UN.[22]

Sustainable land management can benefit every kind of ecosystem, including forests, farmlands, cities, wetlands and oceans, but people have to be persuaded to use them. The European Commission cites a wide range of techniques surveyed in 363 WOCAT studies from across the world (46% from Africa, 41% from Asia, 7% from Europe, and the rest from South America and Australia) that looked at people's attitudes to adopting sustainable practices.[23] Means reviewed range from structural measures such as terraces and dams; agronomic measures such as mulching and increasing organic matter in soil; vegetative measures, such as tree planting and hedging; and management measures, such as grazing timing and change of species compositions. The researchers found that 73% of case study participants had a positive or neutral perception of the cost-benefit ratio of SLM in the short term (three years or less), while 97% had a positive or very positive view of the long term. Their main motivations for adopting sustainable practices were potential production increases (24%); increased profits (20%); improved well-being and livelihood (20%); and reduced workload (5%).

Lack of action to address land degradation will increase emissions and reduce carbon sinks and is inconsistent with the emissions reductions required to limit global warming to 1.5°C or 2°C. (p. 55)

The participants in the WOCAT study were not typical in that they were already involved in sustainable land management, and most of them had received financial support. The IPCC presents measures that have been shown to facilitate implementation of practices that avoid, reduce, or reverse land degradation, including tenure reform, tax incentives, payments for ecosystem services, participatory integrated land-use planning, farmer networks and rural advisory services. Land tenure issues are particularly important for women, who often have less formal access to land than men, and less influence over decisions, even if they carry out many of the land management tasks. Women are also likely to worse affected by climate change, because of discrimination regarding land tenure, less access to other capital assets, and cultural practices.

Land degradation and climate change can therefore act as threat multipliers for people already suffering discrimination and whose livelihoods are precarious, leaving them highly sensitive to extreme climatic events, with consequences such as poverty and food insecurity. In many instances, young people suffer similar disadvantages to women.

Delayed action increases the costs of addressing land degradation, and can lead to irreversible biophysical and human outcomes. Early actions can generate both site-specific and immediate benefits to communities affected by land degradation, and contribute to long-term global benefits through climate change mitigation.

Conclusion

The IPCC makes a strong case for looking at land degradation in terms of the human and natural systems involved, and how they affect and are affected by climate change. Land degradation can be reversed, given the right combination of policies, behaviours and attitudes. In the next chapter, we will look at desertification, partly a natural phenomenon, but also a warning of the extreme consequences of inaction or poorly conceived interventions.

Notes

1. "Rainfall follows the plow", *Encyclopedia of the Great Plains*, David J. Wishart (ed), University of Nebraska–Lincoln, http://plainshumanities.unl.edu/encyclopedia/doc/egp.ii.049

2. UN University *The Value of Land: Prosperous lands and positive rewards through sustainable land management*, Economics of Land Degradation report, 2015

3. IUCN, *Land degradation and climate change: The multiple benefits of sustainable land management in the drylands*, Issues Brief, November 2015 https://www.iucn.org/resources/issues-briefs/land-degradation-and-climate-change

4. IPBES, *The Assessment Report on Land Degradation and Restoration*, 2018 https://www.ipbes.net/sites/default/files/2018_ldr_full_report_book_v4_pages.pdf

5. *Fiscal incentives for agricultural commodity production: Options to forge compatibility with REDD+*, UN-REDD Policy Brief Issue #07, https://www.unep.org/resources/report/fiscal-incentives-agricultural-commodity-production-options-forge-compatibility

6. *A multi-billion-dollar opportunity – Repurposing agricultural support to transform food systems*, FAO, UNDP and UNEP, 2021 https://doi.org/10.4060/cb6562en

7. *Land use and climate change impacts on global soil erosion by water (2015-2070)*, PNAS, Sep 2020, 117 (36) 21994-22001 https://www.pnas.org/content/117/36/21994

8. *Global warming is melting snow faster in the Himalayas*, Deepa Padmanaban, India Climate Dialogue, March 4, 2021 https://indiaclimatedialogue.net/2021/03/04/global-warming-is-melting-snow-faster-in-the-himalayas/

9. "Anthropogenic influences on major tropical cyclone events". Patricola CM, Wehner MF., *Nature* 563, 339-346, 2018, https://www.nature.com/articles/s41586-018-0673-2

10. CGIAR Research Program on Climate Change, Agriculture and Food Security, https://www.cgiar.org/research/program-platform/climate-change-agriculture-and-food-security/

11. "How much carbon can the world's forests absorb? Peter Reich, *The Conversation*, 12 June 2013 https://phys.org/news/2013-06-carbon-world-forests-absorb.html

12. "Amazonia as a carbon source linked to deforestation and climate change", Luciana V. Gatti et al., Nature volume 595, pages388–393 (2021), https://doi.org/10.1038/s41586-021-03629-6

13. UNDP, *Zimbabwe Human Development Report 2017: Climate Change and Human Development*, http://hdr.undp.org/sites/default/files/reports/2842/undp_zw_2017zhdr_full.pdf

14. "Conserving the Cerrado and Amazon biomes of Brazil protects the soy economy from damaging warming", Rafaele Flach et al., *World Development*, Volume 146, October 2021, 105582, https://doi.org/10.1016/j.worlddev.2021.105582

15. "Nonfarm income diversification and household livelihood strategies in Rural Africa: Concepts, dynamics and policy implications", CB Barrett et al., Food Policy, 26 (4) (2001), pp. 315-331, 10.1016/S0306-9192(01)00014-8

16. "Population Pressure: Land and Water: Cars and People Compete for Grain", Lester Brown, in *Plan B 4.0: Mobilizing to Save Civilization*, Earth Resources Institute, 2010, http://www.earth-policy.org/books/pb4/PB4ch2_ss6

17. *Directive 2009/28/EC of the European Parliament and of the Council*
https://eur-lex.europa.eu/LexUriServ/LexUriServ.do?uri=OJ:L:2009:140:001
6:0062:en:PDF

18. "Bioenergy 'flaw' under EU renewable target could raise emissions", Prof
Sir John Beddington, CarbonBrief, 19 December 2017,
https://www.carbonbrief.org/guest-post-bioenergy-flaw-under-eu-renewable-
target-could-raise-emissions

19. Scientist Letter to Biden, von der Leyen, Michel, Suga & Moon Re.
Forest Biomass (February 11, 2021).pdf (dropbox.com)

20. *Bioenergy*, IUCN, 2018,
https://www.iucn.org/renewables-and-biodiversity/our-work/bioenergy

21. UNDER https://www.unccd.int/sites/default/files/inline-files/210601_
RestoringLifetotheLand_Summary2P.pdf

22. *What is ecosystem restoration?*, UNDER, UNEP, FAO, 2021,
https://www.decadeonrestoration.org/what-ecosystem-restoration

23. Markus Giger et al., "Economic Benefits and Costs of Sustainable Land
Management Technologies: An Analysis of WOCAT's Global Data", *Land
Degradation & Development*, 20 August 2015
https://doi.org/10.1002/ldr.2429

Chapter Three
Desertification

In Tacitus' *Agricola*, the Caledonian chieftain Calgacus tells his warriors what he thinks of the Romans who have invaded Britain: "To ravage, to slaughter, to usurp under false titles, they call empire, and where they make a desert, they call it peace".[1] Deserts are a powerful metaphor in many literatures and languages. They can symbolise evil and hell, as here in the mouth of the barbarian leader, or a place for meditation, holiness and self-discovery, as in the stories of Moses in sacred texts or the stories of mystics seeking enlightenment. The difference is one of choice. The positive view accompanies a decision to go to the desert, or circumstances where the stay is only temporary.

Today, some people choose to go and live in the desert, or not to leave it if they were born there, but for the vast majority, the desert is something that occurs without their choosing, even if their actions contribute to its arrival. The consequences can be as devastating as the military invasion Calgacus denounced.

In this chapter, we will look at what causes desertification, and how these causes interact with climate change, usually for the worse.

Desertification is land degradation in arid, semi-arid, and dry sub-humid areas, collectively known as drylands, resulting from many factors, including human activities and climatic variations. The range and intensity of desertification have increased in some dryland areas over the past several decades. (IPCC SRCCL, p. 50)

Drylands cover nearly half the Earth's land area and are home to 3 billion people. As this suggests, the definition of dryland is very broad, covering a range from the Atacama Desert in Chile, sometimes called the driest place on the planet,[2] to a city like Dallas, Texas. Drylands are defined by an aridity index (AI) – the ratio of the amount of precipitation an area receives annually to potential evapotranspiration (the amount of water that the area could return to the atmosphere if there were no limits on supply). Drylands are lands with an AI of less than 0.65, meaning they retain two-thirds or less of precipitation.[3]

Drylands are projected to increase by 11% to 23% by 2100 compared to 1961-90, depending on the success of efforts to reduce emissions. This would mean drylands could make up 50% to 56% of the Earth's land surface by the end of this century, up from around 38% today.[4]

Until the mid-20th century, desertification was attributed almost entirely to human activity, but today the consensus favours an approach highlighting the interactions between human and ecological factors.[5] Archaeologist David Wright applied a systems approach to suggest how the Sahara Desert was created.[6] Wright goes beyond social and environmental systems to include the solar system. Periods of humidity and aridity followed each other throughout the region's history, due to wobbles in the Earth's axis changing the amount of solar radiation the region received. Radiation fluctuations influence rainfall, and during so-called African Humid Periods, north Africa received enough rain to sustain lakes, rivers, grasslands and forests. But after the last humid period around 8000 years ago, the transition to dryland occurred much faster than expected from previous episodes.

The shift to desert would have happened anyway, but it is possible that the regions where agriculture was starting to be adopted in northern Africa were on the verge of ecological regime shifts. Increased pastoralism could have accelerated the process as goats and other animals consumed the vegetation, reducing the amount of atmospheric moisture that helps to produce clouds, and enhancing albedo. The Sahara is continuing to expand, and, as we noted in the opening chapter, is now about 10% bigger than in 1920, with human activity a significant cause.[7] Despite natural inputs to climate fluctuations, the IPCC concludes that: "It is unequivocal that human influence has warmed the atmosphere, ocean and land".[8]

Attribution of desertification to climate variability and change, and to human activities, varies in space and time. (p. 50)

It is also possible that pastoralism and other forms of agriculture were a response by people who had been hunter-gatherers to changing climate conditions that reduced their traditional sources of food. A given factor can have different impacts in different places, depending on how often it occurs and how long it lasts. High natural climate variability in drylands is a major cause of vegetation changes but does not necessarily imply degradation. A single drought for instance can damage land productivity, but as the recovery from the Dust Bowl shows, productivity can be restored, and there may be no lasting degradation. If the droughts become more frequent and more severe, degradation may follow.

The major human drivers of desertification interacting with climate change are expansion of croplands, unsustainable land management practices and increased pressure on land from population and income growth.

Asia has the most people affected by desertification. There are expanding deserts in China, India, Iran, Mongolia and Pakistan; encroaching sand dunes in Syria; steeply eroded mountain slopes in Nepal; deforestation in Laos; and overgrazing in central Asia.[9] In terms of proportion of land affected, Africa is the worst off, with two-thirds of the continent classed as desert or drylands, and nearly three-fourths of the land is estimated to be degraded. Frequent droughts have been particularly severe in recent years in the Horn of Africa and the Sahel. When one thinks of Latin America and the Caribbean, rainforests come to mind, but a quarter of the region is actually desert and drylands, and desertification and degradation of natural resources seriously affect nearly all its countries.

Europe is affected too. In the areas of Southern, Central and Eastern Europe for which data was available, the EU reports that "8% of the territory, corresponding to about 14 million hectares, showed "very high" and "high sensitivity" to desertification. The affected part increases to more than 40 million hectares if moderate sensitivities are also taken into account".[10]

Desertification has already reduced agricultural productivity and provoked income losses of $42 billion a year.[11] In many dryland areas, spread of invasive plants has led to losses in ecosystem services, while groundwater is being depleted by over-extraction. Unsustainable land management, particularly when coupled with droughts, has contributed to higher dust-storm activity, reducing human well-being in drylands and beyond. Higher intensity of sand storms and dune movements are causing disruption and damage to transportation and solar and wind energy infrastructures.

There is a cost to pay in human health too. The IPCC estimates that dust storms were associated with global cardiopulmonary mortality of about 402,000 people in 2005. Apart from respiratory diseases, the WHO blames desertification for higher threats of malnutrition from reduced food and water supplies; more water- and food-borne diseases that result from poor hygiene and a lack of clean water; and the spread of infectious diseases as populations migrate.

Risks from desertification are projected to increase due to climate change.
(p. 50)

The UN provides a set of scenarios known as Shared Socioeconomic Pathways (SSPs), to describe plausible major global developments that together would lead to different challenges for mitigation and adaptation to climate change.[12] The SSPs are based on five narratives describing different kinds of socio-economic development, including sustainable development (SSP1), regional rivalry (SSP3), inequality (SSP4), and fossil-fuelled development (SSP5). SSP2 is the reference or middle-of-the road Pathway, where global social, economic, and technological trends do not shift markedly from historical patterns.

Desertification exacerbates climate change through several mechanisms such as changes in vegetation cover, sand and dust aerosols and greenhouse gas fluxes.

Even a seemingly small change in a major parameter can have dramatic consequences. For example global exposure to multiple risks associated with warming doubles if global mean temperature rises by 2°C rather than 1.5°C,[13] and exposure to risk doubles again with a 3°C increase in global mean temperature compared with 2°C. For populations vulnerable to poverty, the exposure is an order of magnitude greater in the high poverty and inequality scenarios compared to sustainable socioeconomic development. Asian and African regions have 91-98% of the exposed and vulnerable population, with around half of them living in South Asia. In higher warming scenarios, African regions have a growing proportion of the exposed and vulnerable population, ranging from 7 to 17% at 1.5°C, doubling to 14 %-30% at 2°C and 27%-51% at 3°C.

Desertification and climate change, both individually and in combination, will reduce the provision of dryland ecosystem services and lower ecosystem health, including losses in biodiversity. (p. 50)

Desertification was one of the themes studied by the Millennium Ecosystem Assessment.[14] The Assessment identifies four categories of service.

Provisioning Services covers goods produced or provided by ecosystems, including food, fibre, forage, fuelwood, biochemicals, and fresh water.

Regulating Services is defined as benefits obtained from regulation of ecosystem processes, such as water purification and regulation; pollination and seed dispersal; and climate regulation (locally through vegetation cover and globally through carbon sequestration). Drylands store one third of the world's soil carbon.[15] This carbon is stored more permanently than that stored by humid areas, which tend to have greater microbial activity in their soil, reducing permanence.[16]

Cultural Services are nonmaterial benefits obtained from ecosystems, for example recreation and tourism; cultural identity and diversity; landscapes and heritage values; indigenous knowledge systems; and spiritual, aesthetic, and inspirational services.

Supporting Services maintain the conditions for life on Earth, for example soil development, primary production, and nutrient cycling.

Drylands are a major source of biodiversity, with dryland rangelands supplying about a third of terrestrial biodiversity, most of it underground. Up to 2 billion people living in drylands depend on ecosystem services from rangelands. Climate change will exacerbate the vulnerability of dryland populations to desertification. The combination of pressures coming from climate change and desertification will diminish opportunities to reduce poverty, enhance food and nutritional security, empower women, reduce the burden of disease, and improve access to water and sanitation.

Site and regionally-specific technological solutions, based both on new scientific innovations and indigenous and local knowledge (ILK), are available to avoid, reduce and reverse desertification, simultaneously contributing to climate change mitigation and adaptation. (p. 51)

Two of the most ambitious projects to fight desertification involve planting so-called "green walls" of trees to halt the deserts' advance. The idea was first proposed by the British explorer and environmentalist Richard St. Barbe Baker following a long trip to the Sahara in the 1950s, when he called for a thirty mile wide (48 km) "Green Front" involving twenty-four African nations. In 2007, the African Union endorsed the 'Great Green Wall for the Sahara and the Sahel Initiative' (GGWSSI), The African Union describes the Great Green Wall as "an epic ambition to grow an 8,000km natural wonder of the world across the entire width of Africa", that, once completed in 2030, "will be the largest living structure on the planet, 3 times the size of the Great Barrier Reef". Although the wall had only covered 4% of the planned area by September 2020, 350,000 jobs were created and around $90 million in revenues was generated from 2007 to 2018 through the wall-related activities. Over 220,000 people received training on sustainable production of agro-pastoral and non-timber products to support the shift to more responsible consumption and production. The restored area will sequester over 300 MtCO$_2$ by 2030, roughly 30% of the envisioned target.[17]

China's Great Green Wall, officially known as the Three-North Shelter Forest Program, started in 1978, and is planned to be completed around 2050, when it will be 4,500 kilometres long. According to the Chinese Academy of Sciences, nearly 8 million hectares of windbreak trees have been planted; 336,200 square kilometres of desertification land has been fixed, and more than 10 million hectares of desertified grassland have been protected and restored. The grass and forest coverage in the Loess Plateau, the major focus of the program, has been raised to 59.06%. In addition, 400 million tonnes of silt washed into the Yellow River has been reduced annually.[18] Despite these claims, the Program, like Africa's, has been criticised for some of the choices made, and the unintended consequences of certain decisions. Planting trees in arid regions may make water problems worse if the selected trees soak up groundwater. In addition, the fast-growing tree monoculture reduces biodiversity, and is less resilient to disease. It also appears to be less efficient at absorbing carbon than more diverse woodland.[19]

Massive projects may be part of the answer, but they are not the only answer, or necessarily the best or most cost-effective, in particular if they lack the flexibility to evolve along with knowledge or changing conditions. Managing land sustainably can help avoid, reduce or reverse desertification, and contribute to climate change mitigation and adaptation. Sustainable land management practices include:

- reducing soil tillage and maintaining plant residues to keep soils covered;
- planting trees on degraded lands;
- growing a wider variety of crops;

- applying efficient irrigation methods such as drip irrigation;
- moisture conservation methods such as rainwater harvesting using indigenous and local practices;
- improving rangeland grazing by livestock.

Implementation of Land Degradation Neutrality policies allows populations to avoid, reduce and reverse desertification, thus contributing to climate change adaptation with mitigation co-benefits. (p. 52)

Preventing desertification is better than allowing land to degrade and then attempting to restore it, so there is a growing emphasis on avoiding and reducing land degradation, following the Land Degradation Neutrality (LDN) framework. Land degradation neutrality aims to halt the ongoing loss of healthy land through degradation. LDN creates a target to manage land degradation, through measures to avoid or reduce degradation of land, combined with measures to reverse past degradation. The UN Convention to Combat Desertification defines Land Degradation Neutrality as "a state whereby the amount and quality of land resources necessary to support ecosystem functions and services and enhance food security remain stable or increase within specified temporal and spatial scales and ecosystems".[20]

South Africa for example signed the Convention and intends to achieve land degradation neutrality by 2030. As part of the country's efforts, it carried out a study with UNEP (United Nations Environment Programme) to investigate the economic case for ecosystem management and restoration interventions in the Thukela river basin over 2021-2030 to achieve or exceed land degradation neutrality relative to a 2015 baseline.[21] In the river basin, farming, overgrazing and the spread of invasive alien plants have damaged fragile savannah and grasslands. That has hampered the land's ability to sustain livelihoods and to maintain essential ecosystem services, such as supplying water and trapping carbon. The study found that for each Rand invested in full restoration, there is a return of at least 1.7 Rand. The real figure is almost certainly higher, since this estimate only includes selected ecosystem services, and does not include the local multiplier effects of job creation, nor the intangible aesthetic and cultural values associated with restoring landscapes. The estimate of carbon benefits is also conservative, and does not include the global public good benefits of mitigating carbon emissions, only local benefits.

Conclusion

Desertification has been studied for decades and different drivers of desertification have been described, classified, and are generally understood. However, significant knowledge gaps remain as to the extent and severity of desertification at global, regional, and local scales. And despite numerous studies, consistent indicators for attributing desertification to climatic and/ or human causes are still lacking due to methodological shortcomings.

In the next chapter, we will consider food security, looking at the risks and opportunities that climate change presents to food systems, and considering how mitigation and adaptation can contribute to both human and planetary health.

Notes

1. *The Germany and the Agricola of Tacitus. The Oxford Translation Revised, with Notes*, Project Gutenberg,
https://www.gutenberg.org/files/7524/7524.txt

2. *The Driest place on Earth*, Astrobiology at NASA,
https://astrobiology.nasa.gov/news/the-driest-place-on-earth/

3. *What are drylands?* FAO, Dryland Forestry,
http://www.fao.org/dryland-forestry/background/what-are-drylands/en

4. Huang, J., Yu, H., Guan, X. et al. Accelerated dryland expansion under climate change. *Nature Clim Change* 6, 166–171 (2016).
https://doi.org/10.1038/nclimate2837

5. 2007: "Global desertification: Building a science for Dryland Development", Reynolds, J.F. et al., *Science*, 316, 847–851,
https://science.sciencemag.org/content/316/5826/847.full

6. "Humans as Agents in the Termination of the African Humid Period", David K. Wright, *Frontiers in Earth Science*, Vol. 5, 2017,
https://doi.org/10.3389/feart.2017.00004

7. Thomas, Natalie, and Nigam, Sumant. "Twentieth-Century Climate Change over Africa: Seasonal Hydroclimate Trends and Sahara Desert Expansion". *Journal of Climate* 31 (9). United States: American Meteorological Society.
https://doi.org/10.1175/JCLI-D-17-0187.1.
https://par.nsf.gov/biblio/10055432.

8. *Climate change 2021: The physical science basis* IPCC, 2021,
https://www.ipcc.ch/report/ar6/wg1/downloads/report/IPCC_AR6_WGI_
SPM_final.pdf

9. United Nations Convention to Combat Desertification (UNCDC)
https://www.unccd.int/convention/regions

10. *Desertification in the EU: Background paper*, European Court of Auditors,
June 2018,
https://www.eca.europa.eu/Lists/ECADocuments/bp_desertification/bp_
desertification_en.pdf

11. UN Decade for deserts and the fight against desertification 201-2020,
https://www.un.org/en/events/desertification_decade/whynow.shtml

12. *Shared Socioeconomic Pathways*, United Nations Economic Commission
for Europe (UNECE),
https://unece.org/fileadmin/DAM/energy/se/pdfs/CSE/PATHWAYS/2019/
ws_Consult_14_15.May.2019/supp_doc/SSP2_Overview.pdf

13. "Global exposure and vulnerability to multi-sector development and climate
change hotspots", Edward Byers et al., *Environ. Res. Lett.*, 13, May 2018,
https://iopscience.iop.org/article/10.1088/1748-9326/aabf45/meta#artAbst

14. *Ecosystems and Human Well-being: Desertification Synthesis*, Millennium
Ecosystem Assessment Report, WRI, 2005
http://www.millenniumassessment.org/documents/document.355.aspx.pdf

15. *Dryland ecosystems*, IUCN,
https://www.iucn.org/commissions/commission-ecosystem-management/
our-work/cems-specialist-groups/dryland-ecosystems

16. *The practicality of permanence in the soil carbon market*, Real Agriculture, 25 March 2021,
https://www.realagriculture.com/2021/03/the-practicality-of-permanence-in-the-soil-carbon-market/?__cf_chl_jschl_tk__=pmd_7MPtXJSb9gyx.zFwlBzdV.4KmZG3Lo8taQqMiv8Zk9M-1631530284-0-gqNtZGzNAmWjcnBszQjl

17. *The Great Green Wall: Implementation Status and Way Ahead to 2030*, UN Convention to Combat Desertification, 2020,
https://www.unccd.int/publications/great-green-wall-implementation-status-and-way-ahead-2030

18. "China's Three-North Shelterbelt Forest Program brings forest coverage to 13.57 pct", *Xinhua*, 30 November 2018,
http://www.xinhuanet.com/english/2018-11/30/c_137642405.htm

19. "How China's Growing Deserts Are Choking The Country", Daniel Rechtschaffen, *Forbes*, 18 September 2017,
https://www.forbes.com/sites/danielrechtschaffen/2017/09/18/how-chinas-growing-deserts-are-choking-the-country/?sh=2f7eb075d1b9

20. Guide to the scientific conceptual framework for land degradation neutrality | Knowledge Hub (unccd.int)

21. *The potential costs and benefits of addressing land degradation in the Thukela catchment, KwaZulu-Natal South Africa*, J.K. Turpie et al., NCAVES project report, 2021,
https://seea.un.org/content/knowledge-base

Chapter Four
Food security

There have been predictions that the world will face mass starvation ever since the English economist and priest Thomas Malthus published his famous essays on demography back in 1798. As Malthus himself put it in *An Essay on the principle of population*: "The power of population is so superior to the power of the earth to produce subsistence for man, that premature death must in some shape or other visit the human race."[1] And yet it hasn't happened. The UK's population for example doubled over 1750-1800 (the year in which Malthus published *The present high price of provisions*[2]), and tripled over the next century. This demographic surge couldn't have happened without the interaction of a number of elements.

The expression "agricultural revolution" suggests sudden overthrow of the old systems. But even in Britain, the initial changes to agricultural techniques and practices such as enclosing common land and introducing crop rotation were spread over centuries. What accelerated the pace of change was interaction with the industrial revolution. Improved communications and storage and preservation techniques allowed producers to serve markets far from home. A well-functioning financial system provided capital. And a feedback loop was created whereby improved food supplies supported a bigger population that in turn provided labour for emerging industries and markets for farmers.

Likewise, when looking at the prospects for food production and consumption today, we have to look at the whole picture. That means paying special attention to climate change and how it is likely to interact with other factors that influence food security, and the way severe shocks like the Covid-19 pandemic can influence both food production and people's ability to buy food.

Moderate or severe food insecurity at the global level rose from 22.6% in 2014 to 26.6% in 2019, and even before the Covid-19, the high cost of food had already locked 3 billion people out of healthy diets in every region of the world. As a result of the pandemic, food insecurity rose nearly as much in 2020 as in the previous five years combined, to 30.4% according to the 2021 *State of Food Security and Nutrition in the World* report.[3]

Thus, nearly one in three people in the world did not have access to adequate food in 2020 – an increase of 320 million people in just one year, from 2.05 to 2.37 billion. Nearly 40% of those people – 11.9% of the global population, or almost 928 million – faced food insecurity at severe levels.[4] In Northern America and Europe, where the lowest rates of food insecurity are found, the prevalence of food insecurity increased for the first time since the beginning of Food Insecurity Experience Scale (FIES) data collection in 2014.

The rapid growth in agricultural productivity since the 1960s has underpinned the development of the current global food system that is both a major driver of climate change, and increasingly vulnerable to it (from production, transport, and market activities). (IPCC SRCCL, p. 441)

World population has more than doubled since 1960, when it was around 3 billion, compared with over 7.5 billion today. The amount of food available per person though has also more than doubled. Since 1961, food supply per capita has increased more than 30%, accompanied by greater use of nitrogen fertilizers (increase of about 800%) and water resources for irrigation (increase of more than 100%). The food system is under pressure from non-climate stressors such as conflict, population and income growth, and demand for animal-sourced products, as well as from climate change. These stresses are impacting the four pillars of food security: availability, access, utilisation, and stability.

Observed climate change is already affecting food security through increasing temperatures, changing precipitation patterns, and greater frequency of some extreme events. (p. 56)

Climate change poses challenges on a different scale from the variations that can affect crops and livestock during the course of a season or even a year or two. Future changes in the climate could have significant impacts on land use, commodity production, and where different activities are viable.[5] *The State of Food Security* report notes that the number of low and middle-income countries exposed to climate extremes rose from 76% of countries in 2000-2004 to 98% in 2015-2020. Countries' exposure to climate extremes has significantly magnified in terms of frequency, or number of years a country is exposed, from 30% in 2000-2004 to 72% in 2015-2019. In terms of intensity, 52% of countries were exposed to three or four types of climate extremes (heat spell, drought, flood, or storm) in 2015-2020, compared with 11% in 2000-2004. In other words, the number increased nearly five-fold over 2000-2020.

A briefing note prepared for the September 2021 UN Food Systems Summit[6] describes how critical climate variabilities already affect food and nutrition security.[7] Extreme weather events such as heatwaves, droughts and floods impact the productivity of crops, livestock and fisheries. They have an impact on water availability and quality, causing heat stress, altering the pests and disease environment, including the faster spread of mycotoxins and pathogens. Increased frequency and intensity of floods and droughts can lead to harvest failures and infrastructure damage. The exposure of people to heatwaves, droughts and floods can harm their health and lower their productivity, affecting their livelihoods and incomes, especially for those engaged in climate-sensitive sectors or working outdoors. This exposure can strongly affect more vulnerable groups in many lower-income countries, such as smallholder farmers, low earnings households, women and children.

The IPCC quotes research that found that global mean yields of maize, wheat, and soybeans were 4.1, 1.8 and 4.5% lower, respectively, with today's climate, compared to what they would have been if the climate had remained as it was in the pre-industrial era.[8] For rice, no significant impacts were detected, but this study suggests that climate change has modulated recent yields on the global scale and led to production losses. It also suggests that adaptations to date have not been sufficient to offset the negative impacts of climate change, particularly at lower latitudes.

A study in India showed what climate vulnerability means concretely in terms of food security.[9] Researchers examined how the vulnerability of a district to climate can affect child nutrition. They found that districts highly vulnerable to climate change can have more child malnutrition than districts which are relatively less vulnerable. Their analysis shows that the odds of a child suffering from stunting increased by 32%, wasting by 42%, underweight by 45%, and anaemia by 63% if the child belonged to a district categorised as very highly vulnerable when compared to those categorised as very low.

Food security will be increasingly affected by projected future climate change.
(p. 56)

Three of the Shared Socio-economic Pathways (SSPs) mentioned in the previous chapter were used in an assessment of future trends regarding food availability, access, utilisation, and stability: sustainable development (SSP1), regional rivalry (SSP3), and SSP2, where global social, economic, and technological trends do not shift markedly from historical patterns. Global crop and economic models projected a 1 to 29% cereal price increase in 2050 due to climate change, which would impact consumers globally through higher food prices; regional effects will vary. Low-income consumers are particularly at risk, with models projecting increases of 1-183 million additional people at risk of hunger across the SSPs compared to a no climate change scenario. While increased CO_2 is projected to be beneficial for crop productivity at lower temperature increases, it is projected to have lower nutritional quality, with less protein, zinc, and iron.

Historically, the major cause of food insecurity has not been availability but access. Global food production has increased more quickly than population over the past half century, and the EU and USA even had to bring in policies to get rid of "mountains" and "lakes" of food and drink. Access is a different matter: if people are hungry, it is usually because they cannot afford to buy food, not that there is no food to buy. This is not just a problem in developing countries. The prevalence of food insufficiency (low and very low food sufficiency) among U.S. adults rose from 9.5% as of April 23, 2020 to 13.4% as of December 21, 2020, before declining to 8.0% as of April 26, 2021. During this period, very low food sufficiency rose from 2.1% to 3.4%, before declining to 1.9% of US adults as of April 26, 2021. Up to 30.5% of Black households reported that their children sometimes or often did not have enough to eat during the past week as of December 7, 2020, compared to 10.1% for White non-Hispanic households that week.

The models used with the SSPs all predict food price increases by 2050 due to climate change, but the actual increase varies greatly depending on the values of the various socioeconomic and other factors included, ranging from 1 to 29% for cereals for example. Higher prices depress consumer demand, which in turn will not only reduce energy intake (calories) globally, but will also likely lead to less healthy diets with lower availability of key micronutrients and an increase in diet-related mortality in lower and middle-income countries.

A third aspect studied, food utilisation, involves nutrient composition of food, its preparation, and overall state of health. Food safety and quality affects food utilisation. Climate change can change the population dynamics of contaminating organisms due to, for example, changes in temperature and precipitation patterns, humidity, increased frequency and intensity of extreme weather events, and changes in contaminant transport pathways. Changes in food and farming systems, such as intensification to maintain supply under climate change, may also increase vulnerabilities as the climate changes.

> *Agriculture activities within the farm gate and associated land-use dynamics are responsible for about [...] 20% of total anthropogenic emissions.* (p. 476) *Without intervention, these are likely to increase by about 30–40% by 2050, due to increasing demand based on population and income growth and dietary change.* (p. 58)

Food systems emissions also include those beyond the farm gate, such as those upstream from manufacturing of fertilizers, or downstream such as food processing, transport and retail, and food consumption. However, estimating their contribution is very uncertain due to lack of sufficient studies, and estimates of total greenhouse gas emissions attributable to the food system range from about 21 to 37%. It is clear though that the food system is a major contributor to greenhouse gas emissions, but many practices can be optimised and scaled up to advance adaptation throughout the food system.

Enabling conditions need to be created through policies, markets, institutions, and governance, a theme taken up by *Enhancing Nationally Determined Contributions for Food Systems*, a joint study by the UN Environment Programme (UNEP), WWF, EAT, and Climate Focus.[10] Nationally Determined Contributions (NDC) are at the heart of the 2015 Paris Agreement on climate change. In their NDC, each country outlines and communicates their post-2020 climate actions.[11] The report points out that "while many countries mention the agriculture sector in their NDCs, very few set targets in relation to other stages of the food system, such as food loss and waste reduction, sustainable diets or food consumption…and not one country makes reference to food waste…only a handful of NDCs refer to the food system approach, but these mostly remain focused on the stage of food production and not the later stages where large emissions from food loss and waste and diets and consumption occur".

The report's recommendations echo some of those of the IPCC, which argues that supply-side practices can contribute to climate change mitigation by reducing crop and livestock emissions, sequestering carbon in soils and biomass, and by decreasing emissions intensity within sustainable production systems. There is also agreement that on the demand side, consumption of healthy and sustainable diets presents major opportunities for reducing GHG emissions from food systems, and improving health outcomes. Examples of healthy and sustainable include diets that are high in coarse grains, pulses, fruits and vegetables, and nuts and seeds; low in energy-intensive animal-sourced and discretionary foods (such as sugary beverages); and with a carbohydrate threshold.

A change in diet could help address climate change because agricultural activities emit substantial amounts of greenhouse gases. Food supply chain activities past the farm gate also emit GHGs, for instance due to energy used to transport food or keep it refrigerated. GHG emissions from food production vary across food types. Producing animal-sourced food like meat and dairy emits larger amount of GHGs than growing crops, especially in intensive, industrial livestock systems. This is mainly true for commodities produced by ruminant livestock such as cattle, due to enteric fermentation processes that are large emitters of methane. Changing diets towards a lower share of animal-sourced food, once implemented at scale, reduces the need to raise livestock and changes crop production from animal feed to human food. This reduces the need for agricultural land compared to present and thus generates changes in the current food system. From field to consumer this would reduce overall GHG emissions.

Combined supply and demand-side measures can enable the implementation of large-scale land-based adaptation and mitigation strategies, without threatening food security from increased competition for land and higher food prices. The impact of these measures would be multiplied if they were combined with efforts to reduce food loss and waste. The UN created an International Day of Awareness on Food Loss and Waste Reduction to show how reducing food waste and food loss would have multiple benefits.[12] Food loss and waste is responsible for about 7% of global greenhouse gas (GHG) emissions and nearly 30% of the world's agricultural land is currently occupied to produce food that is ultimately never consumed. Nearly 40% of total energy consumption in the global food system is utilized to produce food that is either lost or wasted.

In the US, as much as 40% of all food in the country is going to waste according to the Natural Resources Defense Council, mainly because it is thrown away.[13] Apart from the economic cost of over $2000 per household per year, it is also a problem for climate change. Most of the uneaten food ends up in landfills, where organic matter accounts for 16% of US methane emissions. The EU is the second worse region. The European Commission estimates that food waste per person produces a carbon footprint that is approximately equal to one-third of an individual's average household electricity usage.

In developing countries, due to a warmer climate and a lack of technology, expertise and infrastructure, up to 40% of food can also be wasted, with much of this waste being fresh produce.[14] This is because the farmers are unable to insulate and cool or refrigerate produce after it is harvested, and on the journey between the farm and the consumer, the food can become spoiled. Upgrading infrastructure is one of the main points of a study by the International Food Policy Research Institute. IFPRI projections suggest that without further investment, an additional 78 million people will face chronic hunger in 2050 relative to a no-climate-change future, over half of them in Africa south of the Sahara. IFPRI shows that to offset these impacts would require annual investment in international agricultural research to increase from $1.62 billion to $2.77 billion per year between 2015 and 2050. Additional water and infrastructure investments are estimated to be more expensive than agricultural R&D at about $12.7 billion and $10.8 billion per year, respectively, but these address key gaps to support transformation toward food system resiliency.

A particular concern in regard to the future of food security is the potential for the impacts of increasing climate extremes on food production to contribute to multi-factored complex events such as food price spikes. (p. 514)

In a traditional approach to food security, the market smoothly links demand and supply through prices. This system is essentially in equilibrium until hit by a shock, and the goal is then to get back to normal as quickly as possible. This would be the case in particular if the food system could be assumed to be self-contained, but the Covid pandemic has shown us that the systems we depend on in our daily lives are subsystems of a system of systems, interacting with each other and changing each other through that interaction. A shock to one system can quickly spread, causing failures to cascade, from health to the economy and society for instance.

Given the potential for shocks driven by changing patterns of extreme weather to increase with climate change, there is the potential for market volatility to disrupt food supply through creating food price spikes. This potential is exacerbated by the interconnectedness of the food system with other sectors such as water, energy, and transport, so the impact of shocks can propagate across sectors and geographies. There is also less spare land globally than there has been in the past, such that if prices spike, there are fewer options to bring new production onstream.

Increasing extreme weather events can disrupt production and transport logistics. For example, in 2016, a record yield loss in France is attributed to a conjunction of an abnormally warm late autumn and abnormally high rainfall in the following spring. To the extent that such supply shocks are associated with climate change, they may become more frequent and contribute to greater instability in agricultural markets in the future. Furthermore, conditions similar to past extremes might create significantly greater impacts in a warmer world. A study simulating analogous conditions to the Dust Bowl drought in today's agriculture suggests that Dust Bowl-type droughts today would have unprecedented consequences, with yield losses about 50% larger than the severe US drought of 2012 when corn yield declined 16% compared to 2011 and 25% compared to 2009.[15] Damages at these extremes are highly sensitive to temperature, worsening by about 25% with each degree centigrade of warming. By 2050, over 80% of US summers are projected to have average temperatures that are likely to exceed the hottest summer in the Dust Bowl years.

The principal way in which a shortfall in production or an interruption in trade creates impacts is by affecting agricultural commodity markets. The markets respond to a perturbation through multiple routes, including pressures from other sectors, for example if biofuels policy is encouraging farmers to grow crops for the production of ethanol, as happened in 2007-2008. The market response can be amplified by poor policies, such as countries seeking to ensure their local food security by setting up trade and non-trade barriers to exports. Furthermore, the perception of risks can fuel panic buying on the markets that in turn drives up prices.

Thus, the impact of an extreme weather event on markets has both a *trigger* component (the event) and a risk *perception* component. Through commodity markets, prices change across the world because almost every country depends, to some extent, on trade to fulfil local needs. There is, therefore, a need to build resilience into international trade as well as local supplies.

Conclusion

Food security is influenced by a wide range of interrelated factors, and concentrating on only one of them such as population growth is likely to be misleading as to how to improve the current situation or prepare for the future. **Climate change and its interaction with land does however deserve special attention, since the environment is the system on which all the others ultimately depend.**

In the following chapter, we will look at the interlinkages between desertification, land degradation, food security and GHG fluxes: synergies, trade-offs and integrated response options.

Notes

1. An Essay on the Principle of Population; or A View of its past and present Effects on Human Happiness; With an Inquiry into our Prospects respecting the future Removal or Mitigation of the Evils which it occasions, Thomas Malthus, London,
https://assets.cambridge.org/97805214/19543/
frontmatter/9780521419543_frontmatter.pdf

2. An Investigation of the Cause of the Present High Price of Provisions. By the Author of the Essay on the Principle of Population, Thomas Malthus, London, J. Johnson, 1800,
https://en.wikisource.org/wiki/An_Investigation_of_the_Cause_of_the_
Present_High_Price_of_Provisions

3. The State of Food Security and Nutrition in the World 2021. Transforming food systems for food security, improved nutrition and affordable healthy diets for all, FAO, IFAD, UNICEF, WFP and WHO, 2021, Rome, FAO,
https://doi.org/10.4060/cb4474en

4. The State of Food Security and Nutrition in the World 2021. Transforming food systems for food security, improved nutrition and affordable healthy diets for all, FAO, IFAD, UNICEF, WFP and WHO, 2021, Rome, FAO,
https://doi.org/10.4060/cb4474en

5. *Climate Change, Water and Agriculture: Towards Resilient Systems*, OECD, 2014,
https://read.oecd-ilibrary.org/agriculture-and-food/climate-change-water-and-agriculture_9789264209138-en#page1

6. https://www.un.org/en/food-systems-summit

7. *Climate change and food systems*, Food Systems Summit Brief Prepared by Research Partners of the Scientific Group for the Food Systems Summit, May 2021

https://bonndoc.ulb.uni-bonn.de/xmlui/handle/20.500.11811/9206

8. 2018: "Crop production losses associated with anthropogenic climate change for 1981–2010 compared with preindustrial levels". T. Iizumi, et al., *Int. J. Climatol.*, 38, 5405–5417,
https://doi.org/10.1002/joc.5818 .

9. "Vulnerability of agriculture to climate change increases the risk of child malnutrition: Evidence from a large-scale observational study in India", Bidhubhusan Mahapatra et al., PLOS One, 2021,
https://www.ifpri.org/publication/vulnerability-agriculture-climate-change-increases-risk-child-malnutrition-evidence

10. *Enhancing Nationally Determined Contributions for Food Systems: Recommendations for Decision-makers*, Ingrid Schulte et al., WWF, 2020,
https://wedocs.unep.org/bitstream/handle/20.500.11822/33597/ndcf.pdf?sequence=1&isAllowed=y

11. *Nationally Determined Contributions*, UN Climate Change,
https://unfccc.int/process-and-meetings/the-paris-agreement/nationally-determined-contributions-ndcs/nationally-determined-contributions-ndcs

12. International Day of Awareness on Food Loss and Waste Reduction 29 September
https://www.un.org/en/observances/end-food-waste-day/background

13. *Wasted: How America is losing up to 40 percent of its food from farm to fork to landfill*, NRDC, 2017
https://www.nrdc.org/sites/default/files/wasted-2017-report.pdf

14. *Reducing food waste in developing countries*, University of Sheffield,
https://www.sheffield.ac.uk/sustainable-food/research/reducing-food-waste-developing-countries

15. 2016: "Simulating US agriculture in a modern Dust Bowl drought", M. Glotter and J. Elliott, Nat. Plants, 3, 16193,
https://pubmed.ncbi.nlm.nih.gov/27941818/

Chapter Five
Interlinkages between climate, land and food security

In March 2010, the Eyjafjallajökull volcano in Iceland started erupting, and went on doing so until June. One of the consequences was that in April, thousands of horticultural workers in Kenya lost their jobs. The ash clouds represented a danger to air traffic, and planes were grounded, including the cargo transporters that carried Kenyan flowers to the European markets, causing daily losses of $2 million to flower growers.[1] Like the Covid pandemic, this is an illustration of how in today's interconnected world a shock to one system can have dramatic consequences on other systems that at first sight do not seem interlinked. Global linkages across borders help the world to grow richer, but they also facilitate the spread of shocks among natural and human systems.

Another characteristic of the complex systems we rely on is uncertainty. In its Fifth Assessment Report in 2014, the IPCC acknowledges that there are uncertainties that we will never know and that the best response is to understand and cope with them.[2] The question is how. Conventional risk management tries to stop the problem happening. But however big your flood barrier, sooner or later, water will find a way over, under or around it. So a new approach is emerging from the study of complex systems, based on reinforcing resilience. A resilience approach accepts that all systems might fail. This approach focuses on the ability of a system to absorb, recover from, and adapt to a wide array of shocks to help individuals, communities, and larger groupings not just to deal with adversity, but to adapt to change in a positive way and take advantage of the opportunities it offers.[3]

We are seeing a shift towards a resilience-based approach in at least the discussions around environmental challenges in a number of countries, in Australia for instance as a result of wildfires,[4] or in England following severe flooding.[5] Unfortunately, it usually takes a catastrophe to change people's minds. But people everywhere see that the way humans have changed the environment is having an impact on their daily lives, and usually for the worse. However, as Slovenia's Ambassador to the OECD put it, "we can't, or rather shouldn't, make policy on the basis of anecdotes, however powerful".[6]

This chapter therefore focuses on the interlinkages between options for climate mitigation and adaptation, to address desertification and land degradation, and to enhance food security. The IPCC has identified and fully assessed 40 options.[7] In the following, we will focus on examples concerning land management.

The applicability and efficacy of response options are region context specific; while many value chain and risk management options are potentially broadly applicable, many land management options are applicable on less than 50% of the ice-free land surface. (p. 61) *Landbased options that help mitigate climate change [...] with moderate-tolarge mitigation potential, and no adverse side effects, include options that decrease pressure on land (e.g., by reducing the land needed for food production).* (IPCC SRCCL, p. 646)

Other options in this category include those that help to maintain or increase carbon stores both above-ground (forest measures, agroforestry, fire management, for instance) and below-ground (such as increased soil organic matter or reduced losses; cropland and grazing land management; urban land management; and reduced deforestation and forest degradation). These options also have co-benefits for adaptation by improving health, increasing crop yields, flood attenuation and reducing urban heat island (UHI) effects.

Tackling urban heat islands is a good example. Our impression that hot weather is harder to bear in the city is backed by data. The temperature difference between urban centres and surrounding rural areas ranges from 2°C to 12°C.[8] The heat island effect can exacerbate health impacts by affecting rainfall patterns, interacting with and worsening air pollution, and increasing flood risk and decreasing water quality, but the most direct on human health is through exposure to increased temperature.[9] The 2003 European heatwave killed over 70,000 people, with cities such as Paris particularly hard hit.[10] Energy consumption is also impacted, with one study of Phoenix, Arizona, a semi-arid metropolitan area, showing that air conditioning accounted for around 53% of daytime averaged total electricity demand, ranging from around a third in the morning to two-thirds in the evening.[11]

Numerous technologies are being applied and studied to combat heat islands, surfaces that reflect sunlight back for instance, but one of the most promising ones, and one that has many co-benefits, involves planting vegetation. So-called green infrastructure cools the urban environment through providing shade and evapotranspiration. Research found that "Typically, greenery on the ground reduces peak surface temperature by 2–9 °C, while green roofs and green walls reduce surface temperature by ~17 °C, also providing added thermal insulation for the building envelope".[12] The authors caution however that the cooling potential varies markedly, depending on scale (city or building), greenery extent (park shape and size), plant selection, and plant placement.

The aim of another group of practices is to reduce greenhouse gas emission sources, such as livestock management or nitrogen fertilization management. In Australia, for instance, direct livestock emissions account for about 70% of greenhouse gas emissions by the agricultural sector and 11% of total national greenhouse gas emissions. This makes Australia's livestock the third largest source of greenhouse gas emissions after the energy and transport sectors.[13]

The government describes four main approaches to mitigating livestock greenhouse gas emissions: husbandry (animal breeding, feed supplements, improved pastures); management systems (stocking rates, biological control); numbers of livestock; and manure management. It argues that "Techniques to reduce livestock greenhouse gas emissions may also increase livestock productivity and resilience". For example, measures to change enteric fermentation to reduce emissions may also increase animal productivity by increasing digestive efficiency. But the government experts also warn about possible trade-offs. For example, some methods for reducing livestock emissions per animal may lead to increased dry matter intake per animal or provide the farmer with an opportunity to increase stock, resulting in either no net change or even a net increase in methane production.

> *Land-based options delivering climate change adaptation may be structural (e.g., irrigation and drainage systems, flood and landslide control), technological (e.g., new adapted crop varieties, changing planting zones and dates, using climate forecasts), or socio-economic and institutional (e.g., regulation of land use, associativity between farmers).* (p. 646)

The most efficient strategies will combine structural, technological and socioeconomic and institutional approaches to help communities at different levels adapt rather than simply react to climate change. Africa is expected to see some of the most serious impacts of climate change, and the UN warns that "Taking reactive approaches to food security and disaster recovery costs the people of Africa billions of dollars in lost GDP, and syphons off government resources that should be dedicated to education, social programmes, healthcare, business development and employment".[14]

The UN emphasizes the importance of understanding and addressing system characteristics in climate change policy. Tipping points are particularly important. A system may change slowly and predictably over a long period, then quickly collapse or reconfigure drastically. Trigger points may be reached in Africa if the average global temperature rise is above 2°C and communities do not develop the adaptive capacities required for long-term resilience, provoking an increase in eco-migrants, a rise in food insecurity, more catastrophic disease outbreaks, and increased instability in the region. Climate change multiplies all the risks for Africa's people, environment and stability.

While initial climate change adaptation initiatives show good potential as to economic viability, livelihood enhancement and vulnerability reduction in Africa, long-term sustainability will depend on the prevailing levels of poverty, the wider context of policies and regulations, access to markets and financial services, as well as government capacity to provide continuous technical support to communities. Even so, zero risk is impossible, and helping communities bear the costs of losses will be critical and have to become a bigger part of overall adaptation efforts.

Some adaptation options (e.g., new planting zones, irrigation) may have adverse side effects for biodiversity and water. (p. 646)

Unintended consequences, positive or negative, are another feature of complex systems. Irrigation for example has been shown to have a strong cooling effect during warm extremes in intensely irrigated regions in Southern Europe, North Africa, South Asia, and the United States. Irrigation reduces the probability of hot days by a similar magnitude as global warming increases their likelihood, leading to little or no overall change.[15] The Aral Sea is probably the most infamous example of negative unintended consequences of irrigation.[16] In the 1960s, the Aral was the fourth-largest body of inland water on Earth, but almost all its water vanished after the Soviet government diverted water feeding into the Aral Sea to irrigate farmland in Kazakhstan, Uzbekistan and Turkmenistan. The Aral's water became saltier, concentrations of minerals increased, and fertilizers and pesticides caused pollution. The fishing industry, which had employed about 60,000 people in the early 1960s, disappeared by the 1980s. The dusty lake bed is exposed to the region's high winds, which carry away the salt-laden and polluted soil, creating public health issues. This salty dust is also deposited on cropland, reducing crop yields. The loss of such a substantial body of water has made the region's climate more extreme, with colder winters and hotter and drier summers.

> *Nine options deliver medium-to-large benefits for all five land challenges [...] increased food productivity, improved cropland management, improved grazing land management, improved livestock management, agroforestry, forest management, increased soil organic carbon content, fire management and reduced post-harvest losses.* (p. 61)

Fire management is receiving increasing attention, due to the economic, environmental and health costs of wildfires. For many people, the scale of the problem was revealed by images of large areas of the Arctic burning in 2020, and apocalyptic scenes from Australia and the US that same year. Yet despite record wildfire activity in Australia, the Arctic, Brazil, and the US in 2020, very low levels of wildfire activity in Canada and tropical Africa resulted in a below-average year for global fire-related emissions of carbon to the atmosphere, according to the Copernicus Atmosphere Monitoring Service.[17] The global direct cost of wildfires in 2020 was $17 billion, making it the fifth-costliest year for wildfires, behind 2017, 2018, 2015 (major Indonesian fires), and 2010 (major Russian fires). The human cost is considerable too, even if it is hard to calculate due to lack of data in many affected countries. As mentioned in the first chapter, wildfire smoke is implicated in deaths linked to respiratory diseases, with one study estimating that it is likely responsible for 5,000 to 15,000 deaths in an average year in the US, and higher in smokier years like 2018 or 2020.[18]

Complete data for 2021 are not available at the time of writing, but early indications are that it was a particularly bad year. Carbon emissions from wildfires in Siberia for example were already higher by August than in the whole of 2020, the worst year on record.[19]

In Europe, climate projections show marked increases in fire danger in most regions, especially in western central Europe, by expanding the area with moderate fire danger towards north.[20] The European authorities insist that interactions of climate change with vegetation cover and fire regimes should be fully understood and properly considered in fire management, to enable plans and policies to take into account changes in fuel and vegetation type, changes in burning conditions and additional fire risk.

We have to consider not just how systems interact with each other, but how different aspects of a same system may have to be balanced against each other. Resilience and efficiency for example. Extinguishing forest fires as quickly as possible seems like an efficient policy. Likewise, planting fast-growing species makes economic sense. But extinguish a fire too soon, and you leave behind fuel for a future fire. And fast-growing species have not developed the fire resistance of the slower-growing, traditional varieties. Climate change may bring periodic bouts of heavy rainfall, promoting a flourishing of the undergrowth, followed by droughts, drying it out for the next blaze. The forest may be economically efficient for a time, but it is not resilient, as so many recent examples show. Here, as in other systems, short-term economic efficiency has to be balanced against longer-term sustainability.

A further two options, dietary change and reduced food waste, have no global estimates for adaptation but have medium-to-large benefits for all other challenges. [...] Five options have large mitigation potential (over 3 GtCO$_2$eq yr^{-1}) without adverse impacts on the other challenges [...] increased food productivity; reduced deforestation and forest degradation; increased soil organic carbon content; fire management; and reduced post-harvest losses. (p. 61)

Increasing soil carbon stocks removes CO$_2$ from the atmosphere and increases the water-holding capacity of the soil, thereby conferring resilience to climate change and enhancing adaptation capacity. It is a key strategy for addressing both desertification and land degradation. There is some evidence that increased organic matter content improves crop yields and yield stability. Practices to increase soil organic matter stocks vary in their efficacy. For example, the impact of no-till farming and conservation agriculture on soil carbon stocks is often positive, but depends on the amount of crop residues returned to the soil. If practices to increase soil organic carbon stocks reduce yields, landowners may choose to use the land for a different purpose, such as urban or industrial development, and then increase emissions. In this case, food security could also be affected.

As discussed in Chapter Four, approximately one-third of the food produced for human consumption is wasted in post-production operations, mostly due to poor storage management. Post-harvest food losses underlie the food system's failure to equitably enable accessible and affordable food in all countries. Reduced post-harvest food losses can improve food security in developing countries; while food loss in developed countries mostly occurs at the retail and consumer stage. The key drivers for post-harvest waste in developing countries are structural and infrastructure deficiencies. Thus, reducing food waste at the post-harvest stage requires responses that process, preserve and, where appropriate, redistribute food to where it can be consumed immediately

> *The climate change mitigation potential for bioenergy and BECCS [Bioenergy with carbon capture and storage] is large (up to 11 $GtCO_2$ yr^{-1}); however, the effects of bioenergy production on land degradation, food insecurity, water scarcity, GHG emissions, and other environmental goals [...] depend on the scale of deployment, initial land use, land type, bioenergy feedstock, initial carbon stocks, climatic region and management regime.* (p. 62)

According to the International Energy Agency, negative emissions technologies (NETs) are gaining increasing attention because, without negative emissions, also referred to as carbon dioxide removal (CDR), achieving the goals of the 2015 Paris Agreement would require very steep emission reduction curves up to 2050.[21] For example, to achieve the 2°C target without negative emission technologies, the share of low-carbon sources in global electricity generation would have to increase from 30 to

80% between 2016 and 2030. This would require a pace of transformation in the global electricity sector that may not be feasible.

Bioenergy with carbon capture and storage (BECCS), the process of capturing and storing CO_2 from biomass energy generation is one of the most debated negative emissions options, and an example of the difficult choices when climate goals and food security have to be balanced. Land used for biofuel feedstocks is often unavailable to produce food, and many studies report a negative impact of bioenergy production on food security. Therefore, "to be a feasible alternative to fossil fuels, bioenergy production should provide a balance of energy, environmental, and economic benefits, while not competing for food security either directly or indirectly".[22] A story in the *Financial Times* in September 2021 headlined "Diesel vs Donuts" discussed the conflict. The president of the American Bakers Association stated that his members, including donut manufacturers, were facing higher prices for vegetable oil due to oil companies buying up supplies for biofuels.[23]

Interlinkages of bioenergy and BECCS with climate change adaptation, land degradation, desertification, and biodiversity are highly dependent on local factors such as the type of energy crop, management practice, and previous land use. For example, intensive agricultural practices aiming to achieve high crop yields may have significant effects on soil health, including depletion of soil organic matter, accelerating land degradation and desertification. However, biofuel programmes can be beneficial to future adaptation of ecosystems if they are based on low inputs of fossil fuels and chemicals, limited irrigation, heat/drought tolerant species, and using marginal land.

In the end, there is no perfect solution, and political choices have to be made taking into account the trade-offs between different, and sometimes conflicting, interests and goals. For example, the Drax power station in the UK has been converted to burn wood pellets instead of coal. Two 600+ megawatt biomass units, upgraded with carbon capture technology, could deliver 40% of the negative emissions the UK Climate Change Committee indicates will be needed from BECCS for the country to reach net-zero by 2050. BECCS also has the potential to remove 20-70 million tonnes of CO_2 per year in the UK by 2050. At the same time, in 2018, in order to generate 6% of the UK's electricity, Drax burned 7.1 million tons of pellets – more wood than the entire UK produces every year, with 79% of the biomass supply shipped from North America, generating around 600 million tons of CO_2.[24]

Conclusion

Strategies for climate mitigation and adaptation to address desertification and land degradation, and to enhance food security have a better chance of succeeding if they address the issues systemically, and take account of system properties such as interconnectedness. Timing is also important. Delayed action will result in an increased need for response to land challenges and a decreased potential for land-based response options due to climate change and other pressures. Early action, however, also has challenges including technological readiness, upscaling, and institutional barriers.

The next and final chapter looks at risk management and decision making in relation to sustainable development.

Notes

1. "European flight paralysis exacts high price on the Kenyan flower trade", Nick Wadhams, *The Guardian*, 20 April 2010, https://www.theguardian.com/world/2010/apr/20/kenya-flowers-iceland-volcano

2. *Fifth Assessment Report*, IPCC 2014, https://www.ipcc.ch/assessment-report/ar5/

3. Hynes et al., (2020), *Systemic Thinking for Policy Making: The Potential of Systems Analysis for Addressing Global Policy Challenges in the 21st Century*, New Approaches to Economic Challenges, OECD Publishing, Paris, https://doi.org/10.1787/879c4f7a-en

4. *Resilience and regeneration: be part of Australia's bushfire recovery*, Australia.com, https://www.australia.com/en/travel-inspiration/australia-bushfire-recovery.html

5. Flood and Coastal Erosion Risk Management Strategy Action Plan 2021, Environment Agency, England, 2021, https://www.gov.uk/government/publications/national-flood-and-coastal-erosion-risk-management-strategy-for-england-action-plan/flood-and-coastal-erosion-risk-management-strategy-action-plan-2021

6. *A Systems Approach to Environmental Challenges*, Irena Sodin, Opening Remarks, New Approaches to Economic Challenges Conference on Integrative Economics, Paris, 6 March 2020, https://www.oecd.org/naec/integrative-economics/Ambassador-Sodin-Slovenia-session4-Integrative-Economics.pdf

7. *Interlinkages between desertification, land degradation, food security and greenhouse gas fluxes: Synergies, trade-offs and integrated response options*, IPCC, *Climate Change and Land*, Chapter 6, https://www.ipcc.ch/site/assets/uploads/sites/4/2019/11/09_Chapter-6.pdf

8. "Urban heat island", J.A. Voogt, *Encyclopedia of global environmental change*, Vol. 3, 2003

9. "The Urban Heat Island: Implications for Health in a Changing Environment", Clare Heaviside et al., *Current Environmental Health Reports*, September 2017, https://link.springer.com/article/10.1007/s40572-017-0150-3

10. "Death toll exceeded 70,000 in Europe during the summer of 2003/Plus de 70 000 décès en Europe au cours de l'été 2003", Jean-Marie Robine et al., *Comptes Rendus Biologies*, Volume 331, Issue 2, February 2008, Pages 171-178, https://doi.org/10.1016/j.crvi.2007.12.001

11. "Assessing summertime urban air conditioning consumption in a semiarid environment", F. Salamanca et al., *Environ. Res. Lett.* 8 034022, https://iopscience.iop.org/article/10.1088/1748-9326/8/3/034022/meta

12. "Greenery as a mitigation and adaptation strategy to urban heat", N.H. Wong et al., *Nat Rev Earth Environ* 2, 166–181 (2021), https://doi.org/10.1038/s43017-020-00129-5

13. *Reducing livestock greenhouse gas emissions*, Australia Department of Primary Industries and Regional Development, January 2021, https://www.agric.wa.gov.au/climate-change/reducing-livestock-greenhouse-gas-emissions

14. *Climate Change Adaptation in Africa*, UNDP, 2018, https://www.undp.org/sites/g/files/zskgke326/files/undp/library/Climate percent20and percent20Disaster percent20Resilience/Climate percent20Change/CCA-Africa-Final.pdf

15. W. Thiery, et al., "Warming of hot extremes alleviated by expanding irrigation", *Nat Commun* 11, 290 (2020) https://doi.org/10.1038/s41467-019-14075-4

16. *This Central Asian lake is a stark reminder of the impact we have on the planet*, World Economic Forum, 21 February 2020, https://www.weforum.

org/agenda/2020/02/aral-sea-lake-water-nature-human-impact/

17. "Reviewing the horrid global 2020 wildfire season", Jeff Masters, Yale Climate Connections, 4 January 2021, https://yaleclimateconnections. org/2021/01/reviewing-the-horrid-global-2020-wildfire-season/

18. *Managing the growing cost of wildfire,* Marshall Burke et al., Stanford Institute for Policy Research Policy Brief, October 2020, https://siepr. stanford.edu/sites/default/files/publications/PolicyBrief-October2020_0.pdf

19. *The devastating wildfires of 2021 are breaking records and satellites are tracking it all,* Tereza Pultarova, Science.com, 11 August 2021, https://www. space.com/2021-record-wildfire-season-from-space

20. *Adaptation of fire management plans*, European Climate Adaptation Platform Climate-ADAPT, 31 August 2016 https://climate-adapt.eea. europa.eu/metadata/adaptation-options/adaptation-of-fire-management-plans

21. *Deployment of BECCS/U value chains: Technological pathways, policy options and business models*, IEA Bioenergy, June 2020, https://www. ieabioenergy.com/wp-content/uploads/2020/06/Deployment-of-BECCS-Value-Chains-IEA-Bioenergy-Task-40.pdf

22. "Systematic review on effects of bioenergy from edible versus inedible feedstocks on food security", Selena Ahmed et al., NPJ Sci Food. 2021; 5: 9., Published online 4 May 2021. doi: 10.1038/s41538-021-00091-6

23. "Diesel vs doughnuts': new biofuel refineries squeeze US food industry", Emiko Terazano and Justin Jacobs, *Financial Times,* 8 September 2021, https://www.ft.com/content/b5839a04-a06a-49c1-8622-2974cbb9a84a

24. *Bioenergy with Carbon Capture & Storage (BECCS)*, Geoengineering Monitor, Geoengineering Technology Briefing Jan 2021, https://www. geoengineeringmonitor.org/wp-content/uploads/2021/04/beccs.pdf

Chapter Six
Addressing the risks of climate change

In 2018, the OECD's New Approaches to Economic Challenges initiative organised a conference to discuss lessons learned 10 years after the 2008 financial crisis.[1] John Llewellyn, Chief Global Economist at Lehman Brothers when the crisis broke, described how they had a tool called Damocles for predicting financial crises in developing countries. Out of curiosity, they ran it for the United States in 2007. Damocles predicted a crisis. He told the board, but since the model couldn't say when the crisis would arrive, they decided to ignore the warning, since the alternative was to lose market share and potential dividends. He learned later that in the other big financial institutions, the attitude was the same.

When asked what lessons he drew 10 years after, Llewellyn replied that he saw the same worrying behaviour, with decision-makers ignoring the evidence and the warnings of the experts. But now the dangerous issue was climate change.

The common error he identified is to imagine that the climate, like financial markets and other complex systems, behaves in a linear, predictable manner, when in fact such systems are nonlinear. Recognising this changes how one approaches risk management, and the philosophy underlying efforts to design policies to deal with risk.

This chapter reviews risk and uncertainty surrounding land and climate change in that light, along with policy instruments and decision-making that seek to address those risks and uncertainties. We also look at governance practices that combine responses with co-benefits, lessen the socio-economic impacts of climate change and reduce trade-offs, and advance sustainable land management.

> *The complex spatial, cultural and temporal dynamics of risk and uncertainty in relation to land and climate interactions and food security, require a flexible, adaptive, iterative approach to assessing risks, revising decisions and policy instruments.* (IPCC SRCCL, p. 68)

We argued in the opening chapter that we are not living in the world of Newtonian physics where actions cause predictable reactions. We are part of a complex system of natural, sociopolitical, and economic systems that we are constantly reconfiguring and that is constantly affecting us. Complex adaptive systems are subject to emergent, disruptive crises, whose initial cause often appears insignificant at the time and is only identified afterwards. Such complex systems are prone to cascading failure. But physics and other disciplines teach us a somewhat counter-intuitive lesson on managing this: the more you attempt to optimise complex systems, the more unstable they may become. A systems approach also suggests that while attention usually focuses on triggers – pandemics, natural disasters, geopolitical tensions – we also have to think about the stability, resilience, and functionality of systems under any conditions.[2]

When we talk about the properties of systems designed by humans, we have to remember that they are largely the result of their decisions and behaviour. Policy choices that degrade public services, emphasising short-term efficiency over long-term resilience, reduce our capacity to handle crises and leave us vulnerable to the sort of systemic shock initiated by Covid-19, and the shocks climate change will provoke or compound. Systems thinking allows us to identify the key drivers, interactions, and dynamics of the economic, social, and environmental nexus that policy seeks to shape, and to select points of intervention in a selective, adaptive way. Critically, this allows us to emphasise the importance of system resilience to a variety of shocks and stresses, allowing systems to recover from lost functionality and adapt to new realities.

This approach accepts the inherently uncertain, unpredictable, and even random nature of systemic threats and addresses them through building system resilience. Rather than rely solely upon the ability of system operators to prevent, avoid, withstand, and absorb any and all threats, resilience emphasises the importance of recovery and adaptation in the aftermath of disruption. Resilience acknowledges that massive disruptions can and will happen, and it is essential that core systems have the capacity for recovery and adaptation to ensure their survival into the future.[3] And even take advantage of new or revealed opportunities following the crises to improve the system through broader systemic changes.

Systems thinking cannot, however, provide a roadmap to a desired future. In a complex system, there is no equivalent to a map that shows that following a road gets you to a town without having to physically travel the road first.[4] The terrain is constantly changing as people's behaviour and circumstances change and you have to adapt, adopt or abandon policies as the situation evolves.

The experience and dynamics of risk change over time as a result of both human and natural processes. (p. 675)

There is high confidence that climate and land changes pose increased risks at certain periods of life, especially to the very young and ageing populations, as well as sustained risk to those living in poverty. Research from the Oxford Institute of Population Ageing finds that older adults are vulnerable to being trapped in poor environments through lack of mobility, disability and frailty, for instance leaving them less able to leave coastal cities that because of climate change will be increasingly exposed to the double impact of heat and flooding. They are at increased risk of heat-related illnesses, "compounded by living alone, co-morbidities, medication, and are at higher risk of dehydration than young people, due to the physiological changes that occur as part of the ageing process. Consequently any reduction in access to fresh water, during a drought for example, has severe implications for older adults".[5]

The Oxford Institute also highlights the fact that half the global city population will be over 60 by the year 2050, and will depend on technology to compensate for some of the effects of ageing. However, many of the conveniences that address age-related changes are highly energy dependent, and yet have now become an essential part of modern city life. For example cars and elevators increase access and mobility; air conditioning and heating systems increase comfort; advanced lighting and audio systems compensate for declining senses; and advanced medical interventions delay death. As the populations of cities age, dependencies on these technologies will become more entrenched.

> *There is high confidence that policies addressing vicious cycles of poverty,*
> *land degradation and greenhouse gas (GHG) emissions implemented in*
> *a holistic manner can achieve climate-resilient sustainable development.*
> (p. 675)

Sustainable development pathways supported by effective regulation of land use to reduce environmental trade-offs; reduced reliance on traditional biomass; low growth in consumption and limited meat diets; moderate international trade with connected regional markets; and effective GHG mitigation can result in lower food prices, fewer people affected by floods and other climatic disruptions, and increases in forested land.

A policy pathway with limited regulation of land use, low technology development, resource intensive consumption, constrained trade, and ineffective GHG mitigation instruments can result in food price increases, and significant loss of forest.

> *Delaying deep mitigation in other sectors and shifting the burden to*
> *the land sector, increases the risk associated with adverse effects on food*
> *security and ecosystem services.* (p. 675)

The consequences are increased pressure on land with higher risk of mitigation failure and of temperature overshoot, as well as leaving future generations with the burden of mitigation and unabated climate change. Prioritising early decarbonisation with minimal reliance on carbon dioxide removal decreases the risk of mitigation failure.

Indigenous and local knowledge (ILK) can play a key role in understanding climate processes and impacts, adaptation to climate change, sustainable land management (SLM) across different ecosystems, and enhancement of food security. (p. 676)

The Indian ocean tsunami that struck in December 2004 killed over 220,000 people and caused around $10 billion worth of damage. Casualties were so high because of the scale and speed of the tsunami. Waves were 30 metres high on reaching the shore in some places, after travelling across the open ocean at speeds up to 800 km/hour (500 mph), as fast as a jet plane.[6] Many of the dead were tourists, come to spend the end of year holidays at the beach, but there were fears that the disaster had also wiped some of the Earth's last "Stone Age" peoples, who disappeared from the shores of India's Andaman and Nicobar Islands.

And then they reappeared, explaining that they had moved to higher ground when they realised a tsunami was coming. The Onge tribe, for example, have lived on Little Andaman for between 30,000 and 50,000 years and most of them seem to have survived. Their folklore talks of "huge shaking of ground followed by high wall of water", according to Manish Chandi, an environmental protection worker who has studied the tribes and spoke to some Onges after the disaster.[7]

Knowing a tsunami is coming and what to do to save yourself is an example of indigenous and local knowledge (ILK). For Unesco, "Local and indigenous knowledge refers to the understandings, skills and philosophies developed by societies with long histories of interaction with their natural surroundings. For rural and indigenous peoples, local knowledge informs decision-making about fundamental aspects of day-to-day life. This knowledge is integral to a cultural complex that also encompasses language, systems of classification, resource use practices, social interactions, ritual and spirituality".[8]

As the Onge example shows, this knowledge is context-specific, collective, informally transmitted, and multi-functional. It can encompass factual information about the environment and guidance on management of resources and related rights and social behaviour. It can be used in decision-making at various scales and levels. Exchange of experiences with adaptation and mitigation that include ILK is both a requirement and an entry strategy for participatory climate communication and action.

Opportunities exist for integrating of indigenous and local knowledge with other knowledge. The Intergovernmental Science-Policy Platform on Biodiversity and Ecosystem Services (IPBES) for example identifies content areas where attention to ILK was particularly important for questions in applied ecology.[9] These include enriching understandings of nature and its contributions to people, including ecosystem services; assisting in assessing and monitoring ecosystem change; contributing to international targets and scenario development to achieve global goals like the Aichi Biodiversity Targets[10] and the UN Sustainable Development Goals;[11] and generating inclusive and policy-relevant options for people and nature.

In a similar spirit, improvements to sustainable land management could be achieved by involving people in identifying problems (including species decline, habitat loss, land-use change in agriculture, food production and forestry), selection of indicators, collection of climate data, land modelling, and agricultural innovation opportunities. Meaningful participation overcomes barriers by opening up policy and the science surrounding climate and land decisions to inclusive discussion that promotes alternatives.

Empowering women can bolster synergies among household food security and SLM (Sustainable Land Management). (p. 677)

Women play a significant role in agriculture, food security and rural economies globally, forming 43% of the agricultural labour force in developing countries, ranging from 25% in Latin America to nearly 50% in Eastern Asia and Central and South Europe, and 47% in Sub-Saharan Africa. However, women constitute less than 5% of landholders. In only 37% of 161 developing and developed countries do men and women have equal rights to use and control land, and in 59% customary, traditional, and religious practices discriminate against women, even if the law formally grants equal rights.

Gender is a key axis of social inequality that intersects with other systems of power and marginalisation – including race, culture, class/socio-economic status, location, sexuality, and age – to cause unequal experiences of climate change vulnerability and adaptive capacity. Due to patriarchal social structures and gendered ideologies, women may face multiple barriers to participation and decision-making in land-based adaptation and mitigation actions in response to climate change: disproportionate responsibility for unpaid domestic work, including care-giving activities and provision of water and firewood; risk of violence in both public and private spheres, which restricts women's mobility for capacity-building activities and productive work outside the home; less access to credit and financing; lack of organisational social capital, which may help in accessing credit; and lack of ownership of productive assets and resources, including land.

Adaptation options related to land and climate may produce environment and development trade-offs as well as social conflicts and changes with gendered implications. Collective action and agency of women in farming households have led to prevention of crop failure, reduced workload, increased nutritional intake, increased sustainable water management, diversified and increased income and improved strategic planning. Women's waged labour can help stabilise income from more land- and climate-dependent activities such as agriculture, hunting, or fishing.

Integrating, or mainstreaming, gender into land and climate change policy requires assessments of gender-differentiated needs and priorities, selection of appropriate policy instruments to address barriers to women's sustainable land management such as gender considerations as qualifying criteria for funding programmes or access to financing for initiatives. Training and extension programmes for women to facilitate sustainable practices is also important. Such training could be built into existing programmes or structures, such as collective microenterprise or through women's community-based organisations, although not in such a way that it overwhelms these organisations or supersedes their existing missions. Sustainable land management programmes could also benefit from the voluntary engagement of men in gender-equality training and initiatives.

Securing land tenure for women promotes sustainable land management. It increases their conservation efforts, environmentally beneficial agricultural investments such as tree planting and also improves their income. If women had the same access to productive resources as men, the number of hungry people in the world could be reduced by 12 to 17%.[12]

The significant social and political changes required for sustainable land use, reductions in demand and land-based mitigation efforts associated with climate stabilisation require a wide range of governance mechanisms. (p. 70)

There is robust evidence and high agreement that resource and disaster crises are crises of governance. As Amartya Sen said, "There is no such thing as an apolitical food problem".[13] **Adaptive governance of risk combining resilience and complexity theory has emerged in response to these crises and involves four critical pillars: sustainability; governance; mitigation; and adaptation.**

Anticipation is a key component of adaptive climate governance, with steering mechanisms developed to adapt to and shape uncertain futures. Policymakers can operationalise anticipatory governance through:

- a foresight system considering future scenarios and models;
- a networked system for integrating this knowledge into the policy process;
- a feedback system using indicators to gauge performance;
- an open-minded institutional culture allowing for hybrid and polycentric governance.

In order to manage uncertainty, natural resource governance systems need to allow agencies and stakeholders to learn and change over time. It means responding to ecosystem changes and new information with different management strategies and practices that involve experimentation. Experiments in emissions trading between 1968 and 2000 in the US for example helped to realise specific models of governance and material practices through mutually supportive lab experiments and field applications that advanced collective knowledge.

A sustainable land management plan is dynamic and adaptive over time to (unforeseen) future conditions by monitoring indicators as early warnings or signals of tipping points, initiating a process of change in policy pathway before a harmful threshold is reached. This process has been applied to coastal sea level rise, starting with low-risk, low-cost measures and working up to measures requiring greater investment after review and re-evaluation. Viewing climate mitigation as a series of connected decisions over a long time period and not an isolated decision reduces the fragmentation and uncertainty of models and increases the effectiveness of policy measures.

Adaptive governance addresses large uncertainties and their social amplification through different perceptions of risk. It co-evolves with risk by implementing policy mixes and assessing effectiveness in an ongoing process, making corrections when necessary. In respect of climate adaptation to coastal and riverine land erosion due to extreme weather events impacting on communities, adaptive governance offers the capacity to monitor socio-economic processes and implement dynamic locally informed institutional responses.

In comparison to other governance initiatives of ecosystem management aimed at conservation and sustainable use of natural capital, adaptive governance has visible effects on natural capital by monitoring, communicating and responding to ecosystem-wide changes at the landscape level.

Conclusion

Environmental challenges are systemic, complex and nonlinear, and although they will affect the poorest regions and poorest people most, the floods and fires in Europe and the US in 2021 show that climate-related disasters are becoming more deadly everywhere. The rate of climate change has implications for how quickly systems can adapt. Action at a global level will determine the intensity of potential trade-offs between mitigation measures and their possible impacts on ecosystems, and human well-being at smaller scales. Climate-related risk cannot be avoided, and further climate-related shocks or more prolonged stresses are inevitable. Windows of opportunity for redesigning and implementing mitigation and adaptation can arise in the aftermath of a major disaster or extreme climate event. They can also arise when collective action and citizen science motivate voluntary shifts in lifestyles supported by top-down policies.

Notes

1. *10 Years after the failure of Lehman Brothers: What have we learned?*, OECD New Approaches to Economic Challenges Conference, 13-14 September 2018, https://www.oecd.org/naec/10-years-after-the-crisis/

2. *Debate the Issues: Complexity and Policymaking*, Love, P. and J. Stockdale-Otárola (eds.), OECD Insights, OECD Publishing, Paris, https://doi.org/10.1787/9789264271531-en

3. *The Science and Practice of Resilience*, Igor Linkov and Benjamin Trump, Springer, 2019, https://link.springer.com/book/10.1007%2F978-3-030-04565-4

4. *The end of theory*, Richard Bookstaber, Princeton University Press, 2018, https://press.princeton.edu/books/hardcover/9780691169019/the-end-of-theory

5. Harper, S. "The Convergence of Population Ageing with Climate Change", Sarah Harper, *Population Ageing* 12, 401–403 (2019), https://doi.org/10.1007/s12062-019-09255-5

6. *Tsunami Historical Series: Sumatra – 2004*, NOAA, https://sos.noaa.gov/catalog/datasets/tsunami-historical-series-sumatra-2004/

7. *Tsunami folklore 'saved islanders'*, Subir Bhaumik, BBC News, 20 January 2005, http://news.bbc.co.uk/2/hi/south_asia/4181855.stm

8. *What is local and indigenous knowledge?* Unesco, 2017, http://www.unesco.org/new/en/natural-sciences/priority-areas/links/related-information/what-is-local-and-indigenous-knowledge/

9. "Working with Indigenous and local knowledge (ILK) in large-scale ecological assessments: Reviewing the experience of the IPBES Global Assessment", Pamela McElwee et al., *Journal of Applied Ecology*, 28 July

2020, https://doi.org/10.1111/1365-2664.13705

10. *Aichi Biodiversity Targets*, Convention on Biological Diversity, 18 September 2020, https://www.cbd.int/sp/targets/

11. *Seventeen Goals to Transform our World*, United Nations, https://www.un.org/sustainabledevelopment/

12. *Women in Agriculture: Closing the gender gap for development*, FAO, 2010, http://www.fao.org/publications/sofa/2010-11/en/

13. *Poverty and Famines: An Essay on Entitlement and Deprivation*, Amartya Sen, Oxford: Clarendon Press, 1981, https://oxford.universitypressscholarship.com/view/10.1093/0198284632.001.0001/acprof-9780198284635

What to do with this book once you don't want it anymore?

1. Give it a second life

Extend the life span of this book by giving it to someone or to a charity or by reselling it. A book with little damage can still be useful before being thrown away.

2. Recycle it well

If it has to be thrown away, put it in the recycling bin so that the paper it is made of can be reused. But be careful, it is important to remove recycling disruptors first:

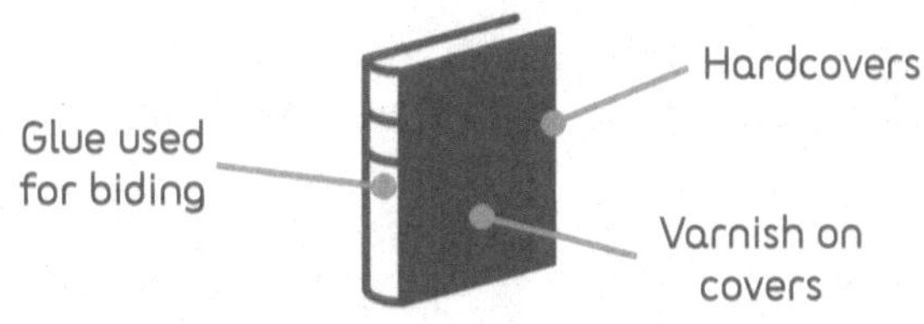

To do this, simply remove its cover and put it in the bin for packagings and all the pages left in the waste paper bin.

Legal deposit: December 2021